2

Masonry Walls

Masonry Walls

Specification and Design

K. Thomas MSc, CEng FIStructE, FCIOB, ARTC

*Principal, Kenneth Thomas Associates, Consulting
Structural Engineers and Building Technologists*

Butterworth-Heinemann Ltd
Linacre House, Jordan Hill, Oxford OX2 8DP

A member of the Reed Elsevier group

OXFORD LONDON BOSTON
MUNICH NEW DELHI SINGAPORE SYDNEY
TOKYO TORONTO WELLINGTON

First published 1996

© K. Thomas 1996

British Library Cataloguing in Publication Data
A catalogue record for this book is available from the British Libraary.

ISBN 0 7506 2465 5

Library of Congress Cataloging in Publication Data
A catalogue record for this book is available from the Library of Congress.

Typeset by Keyword Typesetting Services
Printed in Great Britain by Hartnolls Limited, Bodmin, Cornwall

Contents

Preface

Although masonry has been used as a building material since the beginning of time, some members of the construction industry may be less familiar with many of its specific properties than they are with other building materials and a great deal is often left to chance in specifying, designing and constructing masonry walls.

The purpose of the book is to provide a detailed reference book for construction professionals responsible for specifying and designing masonry structures. It provides detailed information on the units of construction and mortars and general guidance on the selection of materials for the various locations and site exposures.

The reasons for, and accommodation of, movements in masonry are discussed in detail, as well as bricklaying and blocklaying under winter conditions, frost attack, salts and stains, rain penetration, dampness in walls and remedial measures, wall finishes (plastering, rendering and painting etc.) and the thermal and sound insulation of walls.

The book does not cover advanced structural analysis and design, which is adequately covered elsewhere, but does give guidance on the empirical design of freestanding walls, laterally loaded and internal walls and partitions. It also gives an introduction to calculated loadbearing masonry, fin and diaphragm walls, reinforced and post-tensioned walls and masonry cladding to timber framed construction. Fire resistance of masonry is discussed, as well as workmanship, quality control, bonds and finishes, and repairing and replacing masonry.

The author wishes to express his thanks to his many former colleagues and friends at the Brick Development Association; British Ceramic Research Limited; the Building Research Establishment; the Mortar Producers Association and the British Standards Institution for their help and advice over a period of many years which has made the preparation of this book possible. In particular he wishes to thank Dr (Timber) West and Mr Donald Foster for their generous assistance. The author is also grateful to the many brick and block manufacturers, mortar producers and other associated companies who have provided help and assistance.

Grateful thanks are also due to my good friend Mr John Tennant for designing the front cover of the book.

Extracts from British Standards are reproduced with the permission of British Standards Institution, 389 Chiswick High Road, London W4 4AL.

K.T.

1
Bricks and blocks

A wide variety of bricks and blocks is manufactured throughout the world from a range of materials, but in the western world bricks and blocks are normally produced from either fired clay, calcium silicate (sandlime and flintlime) or concrete. Sizes and shapes of bricks and blocks are legion and definitions of brick and block depend upon the country of origin but the British Standard BS 3921: 1985 *Clay bricks* defines a brick as 'a masonry unit not exceeding 337.5 mm in length, 225 mm in thickness (referred to as width in one of the standards) or 112.5 mm in height'. Clay blocks are not currently produced in the UK but BS 6073: Part 1: 1981 *Pre-cast concrete masonry units* defines a block as 'a masonry unit which, when used, in its normal aspect exceeds the length or width or height specified for bricks'.

1.1 Clay bricks

Clay bricks are manufactured from fired clay and have a wide range of different colours and textures. They are produced in many different shapes, sizes and strengths, with differing material properties such as water absorption, suction rate and compressive strength. These properties are determined by the type of clay, kiln and the method of forming the bricks (i.e. by moulding or extrusion).

The colour and texture of clay bricks can be natural (and depend upon the type of clay and method of forming and firing) or a surface colour and/or texture may be applied while the bricks are in the 'green' or unfired state.

The following varieties of clay bricks are available.

Common bricks. Suitable for general building work but having no special claim to give an attractive appearance.

Facing bricks. Specially made or selected to give an attractive appearance when used without rendering or plaster or other surface treatment of the wall.

Engineering bricks. Having a dense and strong semi-vitreous body conforming to defined limits for absorption and strength.

Some non-standard specifications refer to semi-engineering bricks but this is a colloquial term without recognized definition and should never be used by specifiers.

Durability

Clay bricks are classified in terms of durability (i.e. frost resistance and soluble salt content) as follows:

Frost-resistant (F). Bricks durable in all building situations including those where they are in a saturated condition and subjected to repeated freezing and thawing.

Moderately frost-resistant (M). Bricks durable except when in a saturated condition and subjected to repeated freezing and thawing.

Not frost-resistant (O). Bricks liable to be damaged by freezing and thawing if not protected as recommended in BS 5628: Part 3:1985 during construction and afterwards, e.g. by an impermeable cladding. Such bricks may be suitable for internal use.

Low soluble salts (L). Percentage by mass of soluble salts not exceeding the following:

    calcium 0.3 per cent
    magnesium 0.03 per cent
    potassium 0.03 per cent
    sodium 0.03 per cent
    sulphate 0.5 per cent

For normal soluble salts (N). There is no limit on soluble salt content. Based on the foregoing, bricks may have one of the following durability designations:

    FL – frost-resistant (F) and low soluble salt content (L)
    FN – frost-resistant (F) and normal soluble salt content (N)
    ML – moderately frost-resistant (M) and low soluble salt content (L)
    MN – moderately frost-resistant (M) and normal soluble salt content (N)
    OL – not frost-resistant (O) and low soluble salt content (L)
    ON – not frost-resistant (O) and normal soluble salt content (N)

BS 3921: 1985 also specifies that no brick shall show efflorescence worse than moderate, i.e. more than 10 per cent but not more than 50 per cent of the area of the face covered with a deposit of salts (under a prescribed test procedure). Efflorescence is given more detailed consideration in Chapter 6.

Types of brick

Solid brick

A solid brick is a brick having no holes, cavities or depressions in it. Up until the 1985 revision of BS 3921 solid bricks were permitted to have up to 25 per cent of the gross volume perforated but this anomaly was removed in 1985. However, this does not affect recommendations in other documents that a structural brick should be treated as solid, e.g. in respect of its behaviour under compressive loading and in fire reference tables.

Perforated brick
A perforated brick is a brick having holes passing through it which do not exceed 25 per cent of the gross volume. The holes are so arranged that the aggregate thickness of solid material, when measured horizontally across the width of the brick at right angles to the face, is nowhere less than 30 per cent of the overall width of the brick. The area of any one hole must not exceed 10 per cent of the gross area of the brick.

Cellular brick
This is a brick in which cavities (holes closed at one end) exceed 20 per cent of the gross volume.

Frogged brick
A frogged brick is a brick having a depression in one or more bed faces which in total does not exceed 20 per cent of the gross volume of the unit (Figure 1.1).

Dimensions

When specifying bricks the 'work size' should be quoted, that is the face dimensions and thickness. Table 1.1 gives work sizes for the standard brick as well as the co-ordinating dimensions. Table 1.2 gives dimensions for modular bricks. Other formats are available from some manufacturers.

Appearance

Clay bricks are required to be well fired (a difficult term to define) and reasonably free from deep or extensive cracks and from damage to edges and corners, from pebbles and expansive particles of lime.

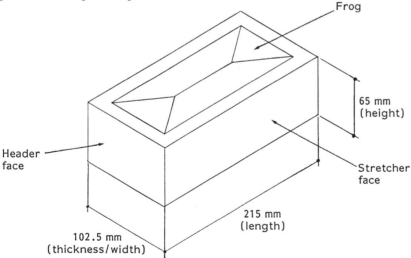

Figure 1.1 Standard size deep frogged clay brick

TABLE 1.1. Dimensions for standard clay brick

Work size + 10 mm	Specification dimensions (mm)	
*Co-ordinating size Length × height 225 × 75	Work size Length × height 215 × 65	Work size Thickness/width 102.5

TABLE 1.2 Dimensions for modular clay bricks

*Co-ordinating size Length × height 200 × 75	Work size Length × height 190 × 65	Work size Thickness/width 90

*Co-ordinating size – this is the size of the space, bounded by co-ordinating planes, allocated to a component, including the allowance for joints which is normally 10 mm

The texture is entirely dependent upon the method of manufacture and the clay used for making the bricks. Some bricks are extremely uniform while others are not, although non-uniformity of both texture and colour is often considered a desirable property.

Compressive strength and absorption

Clay bricks have strength and absorption requirements as indicated in Table 1.3.

TABLE 1.3 Strength and absorption requirements for clay bricks

Designation	Class	Minimum average compressive strength (N/mm^2)	Average absorption (% by weight)
Engineering	A	$\geqslant 70$	$\leqslant 4.5$
Bricks	B	$\geqslant 50$	$\leqslant 7.0$
D.p.c. bricks (1)		$\geqslant 5$	$\leqslant 4.5$
D.p.c. bricks (2)		$\geqslant 5$	$\leqslant 7.0$
Other bricks		$\geqslant 5$	No limits

Note: d.p.c. bricks (1) are recommended for use in buildings but d.p.c. bricks (2) are only for use in external works (see Table 13 of BS 5628: Part 3: 1985).

Absorption

A low water absorption figure is used in defining engineering and damp-proof course (d.p.c.) bricks but water absorption, like strength, is not a general index of durability. With many but not all clays, the more durable bricks absorb less water than those that are less durable.

Initial rate of absorption

The initial rate of absorption (IRA) of a fired clay brick is a function of the pore structure which controls the suction or rate of absorption, due to capillary action. It is important that the IRA should be within certain limits, as too high a value may seriously affect the mortar water content, thus possibly reducing tensile bond strength. Too low an IRA may make the units extremely difficult to lay due to floating action. According to BCRL Special Publication 56 and BS 5628: Part 3 optimum bond is likely to be achieved when the IRA of clay bricks at the time of laying is not greater than $1.5 \, \text{kg/m}^2/\text{min}$ or alternatively a water-retentive mortar is used.

Bricks having a high suction should be wetted so that an acceptable value is achieved prior to laying. The amount of wetting to achieve the suction rate required will vary depending upon the type of brick and its condition when received on site. Guidance on wetting procedures can generally be obtained from the manufacturer of the brick. Excessive wetting does not lead to optimum adhesion and can cause staining of the brickwork.

1.2   Calcium silicate (sandlime and flintlime) bricks

Calcium silicate bricks consist basically of a mixture of sand or flint mixed with lime, which is mechanically pressed together and combined by the action of steam under pressure. When only natural sand is used with the lime the bricks may be described as 'sandlime' bricks. Alternatively, where a substantial proportion of crushed flint is included they may be described as 'flintlime' bricks. The bricks can be of natural colour, or pigments may be included to provide a wide range of colour. The colour of calcium silicate bricks is darker when wet than when dry.

Textured bricks can be produced by special arrangement with some manufacturers.

Types of brick

Solid bricks and bricks with frogs have the same definition as for clay bricks. Cellular bricks and perforated bricks are not produced in the UK.

Dimensions

When specifying dimensions, the 'work size' should be quoted, that is the face dimensions and thickness. Table 1.4 gives work sizes for the standard brick as well as the co-ordinating dimensions and allowable manufacturing tolerances. It should be noted that BS 187: 1978, the standard for calcium silicate (sandlime and flintlime) bricks, adopts a different approach to dimensional accuracy to that of the clay brick standard BS 3921: 1985 and readers are advised to study the appropriate Standards before specifying the units. BS 3921: 1985 bases dimensional deviations on the overall measurements of 24 bricks whereas BS 187: 1978 adopts a tolerance on individual bricks based on a sample of 10 units.

TABLE 1.4 Dimensions based on standard calcium silicate brick

	Length (mm)	Thickness/width (mm)	Height (mm)
*Co-ordinating size	225	112.5	75
Work size	215	102.5	65
Maximum limit of manufacturing size	217	105	67
Minimum limit of manufacturing size	212	101	63

*Co-ordinating size – this is the size of the space, bounded by co-ordinating planes, allocated to a component, including the allowance for joints which is normally 10 mm

Table 1.5 gives dimensions for modular bricks.

TABLE 1.5 Dimensions based on modular calcium siicate bricks

Work size + 10 mm	Specification dimensions (mm)	
Co-ordinating size (mm) Length × height	Work size Length × height	Work size Thickness/width
300 × 100	290 × 90	90
200 × 100	190 × 90	90
300 × 100	290 × 90	65
200 × 100	190 × 90	65

Appearance

Calcium silicate bricks are required to be free from visible cracks and noticeable balls of clay, loam and lime, and the colour and texture should be to the specifier's requirements. These bricks are normally extremely uniform in appearance due to the materials and manufacturing process.

Compressive strength

The compressive strength of the bricks taken for design purposes is the mean compressive strength but BS 187: 1978 also requires a predicted lower limit of compressive strength as shown in Table 1.6.

TABLE 1.6 Compressive strength requirements for calcium silicate bricks

Designation	Class	Compressive strength (N/mm^2)	Predicted lower limit comprehensive strength (N/mm^2)	Strength colour mark
Loadbearing	7	48.5	40.5	Green
or	6	41.5	34.5	Blue
Facing	5	34.5	28.0	Yellow
	4	27.5	21.5	Red
	3	20.5	15.5	Black

The term 'class' in relation to strength is a hangover from the days of imperial measurement when brick compressive strengths were stated in lbf/in^2. 1000 lbf/in^2 would be Class 1, 2000 lbf/in^2 Class 2, etc.

Marking

As each of the different classes of brick will be near identical in colour a proportion of the bricks in each delivery may be colour-marked by the manufacturer to show the strength class. Where this is done the colours used are as given in Table 1.6.

1.3   Concrete bricks

A variety of concrete bricks is produced but for convenience they can be listed under the following headings.

Dense aggregate concrete bricks. These are manufactured by hydraulically pressing a concrete mix into individual preformed moulds. The unit is instantly demoulded onto pallets and steam cured. The bricks may be frogged or not, depending upon the method of manufacture.

Lightweight concrete bricks. These are produced from lightweight aggregate or autoclaved aerated concrete and are made in the same manner as concrete blocks of the same materials, (see p. 8).

Types

Solid brick

BS 6073: Part 1: 1981 defines a solid brick as 'a brick in which small holes passing through, or nearly through the brick do not exceed 25 per cent of its volume, or in which frogs [depressions in the bed faces of a brick] do not exceed 20 per cent of its volume'. Small holes are defined as being less than 20 mm wide or less than 500 mm^2 in area. Up to three larger holes, not exceeding 3250 mm^2 each, are permitted as aids to handling within the total of 25 per cent.

Concrete bricks in the UK are currently manufactured either totally solid or with frogs and the above definition's reference to perforations in solid bricks is therefore of academic interest only.

Perforated, hollow and cellular bricks

These are also defined in BS 6073: Part 1: 1981 but are not currently produced in the UK.

Fixing unit

A fixing unit is a masonry unit of the same dimensions as a brick which permits the easy driving of, and provides a good holding power for, nails and screws.

Dimensions

When specifying, the 'work size' should be quoted, that is the face dimensions and thickness. Table 1.7 gives work sizes normally available, as well as the co-ordinating dimensions.

TABLE 1.7 Dimensions for concrete bricks

Work size + 10 mm	Specification dimensions (mm)	
Co-ordinating size (mm)*	Work size	Work size
Length × height	Length × height	Thickness/width
300 × 100	290 × 90	90
225 × 75	215 × 65	103
200 × 100	190 × 90	90
200 × 75	190 × 65	90

*Co-ordinating size – this is the size of the space, bounded by co-ordinating planes, allocated to a component, including the allowance for joints which is normally 10 mm.

Appearance

Facing bricks are available in a variety of colours and textures and the range is dependent upon the manufacturer. The colour of concrete bricks is darker when wet than when dry.

Compressive strength

BS 6073: Part 1: 1981 requires fixing bricks to have a minimum compressive strength of 2.8 N/mm^2 and other bricks 7.0 N/mm^2. However, dense aggregate concrete bricks normally have a minimum compressive strength of 20 N/mm^2.

1.4   Blocks

Fired clay blocks

Fired clay blocks are used for walling and as an infill for flooring outside the UK, but are no longer produced in the UK and no further reference will be made to this product.

Concrete blocks

A wide variety of concrete blocks is produced in the UK but for convenience they can be listed under three headings.

Dense and lightweight aggregate concrete blocks. The blocks are manufactured from aggregate and cementitious binder, which is compacted into mould boxes and instantly demoulded and cured by either controlled accelerated techniques or, alternatively, by air curing.

Reconstructed stone masonry units. These units are manufactured from aggregate and cementitious binder using casting, or pressing techniques. They are intended to resemble natural stone and to be used for similar purposes.

Autoclaved aerated concrete blocks. The blocks are manufactured from a slurry incorporating cement, finely ground sand/or pulverized fuel ash (PFA), lime and aluminium powder. The mix is cast in large moulds and the resulting 'cake' cut with wires to form blocks which are subsequently autoclaved in superheated steam.

Dimensions

When specifying, the 'work size' should be quoted, that is, the face dimensions and thickness. Table 1.8 gives work sizes normally available, as well as the co-ordinating dimensions.

Types

Solid block
BS 6073: Part 1: 1981 defines a solid block as one which contains no formed holes or cavities other than those inherent in the material.

TABLE 1.8 Dimensions for concrete blocks

Work size + 10 mm	Specification dimensions (mm)	
*Coordinating size	Work size	Work size
Length × height	Length × height	Thickness/width
400 × 200	390 × 190	60, 75, 90, 100, 115, 140, 150, 190, 200
450 × 150	440 × 140	60, 75, 90, 100, 140, 150, 190, 200, 225
450 × 200	440 × 190	60, 75, 90, 100, 140, 150, 190, 215, 220
450 × 225	440 × 215	60, 75, 90, 100, 115, 125, 140, 150, 175, 190, 200, 215, 220, 225, 250
450 × 300	440 × 290	60, 75, 90, 100, 140, 150, 190, 200, 215
600 × 150	590 × 140	75, 90, 100, 140, 150, 190, 200, 215
600 × 200	590 × 190	75, 90, 100, 140, 150, 190, 200, 215
600 × 225	590 × 215	75, 90, 100, 125, 140, 150, 175, 200, 215, 225, 250

*Co-ordinating size – this is the size of the space, bounded by co-ordinating planes, allocated to a component, including the allowance for joints which is normally 10 mm.

Cellular block
A cellular block is one which has one or more formed holes or cavities which do not wholly pass through the block.

Hollow block
A hollow block is one which has one or more formed holes or cavities which pass through the block.

Compressive strength

The following are compressive strengths of concrete blocks commonly used for design purposes:

2.8, 3.5, 5.0, 7.0, 10.0, 15.0, 20.0 and 35 N/mm^2.

Other compressive strengths are available and in use but not all manufacturers necessarily produce the complete range.

1.5   Natural stone

Natural stone falls into three primary classifications or groups:

(a)   igneous or primary, generally granite, basalt, diorite and serpentine;
(b)   sedimentary or secondary, generally limestone and sandstone;
(c)   metamorphic or tertiary, generally marble or slate but there are many subdivisions, combinations and conglomerates.

Processing

Natural stone is processed as follows.

(1)   Quarrying or mining the stone by manual cutting, percussion, hydraulic, blasting and power sawing.

(2)   Seasoning – this varies from quarry to quarry. Many stones can be worked by hand more easily immediately after quarrying but because of the high water content some stone, when mined in winter, is not exposed until the spring. Salts in the stone will tend to move towards the surface during seasoning and are removed when the stone is dressed.

(3)   Preparation can be by hand working or by mechanical means. Most stones are prepared by mechanical primary sawing.

(4)   Finishing is carried out by planing or milling. For marble, granite and some hard limestones there are various degrees of polished finish.

Other surface finishes include those of mechanical or hand tooling to obtain different surface effects. Finally, there is banker mason work for aesthetic or functional effects including profiling to shed rainwater from the building face.

Strength

Structural masonry built with natural stonework is designed on a similar basis to that constructed in solid concrete blocks. Where the masonry is formed from large, carefully shaped pieces with relatively thin joints, its loadbearing capacity is more closely related to the intrinsic strength of the stone than when small stones are used. Suppliers should be consulted for more detailed information on ultimate strength for use in conjunction with BS 5628: Part 1: 1978.

The strength of random rubble masonry is normally taken as 75 per cent of the corresponding strength of natural stone masonry.

Selection of stone

The selection of stone is liable to be affected by variation in bed strata. In some quarries the beds are sufficiently regular for selection to present no difficulty, the relative durability of stone from such quarries being fairly well known, whereas in others, variations are such that each individual block needs to be considered separately. The quality of such blocks is not always apparent and local knowledge is usually the best guide. It is advisable to take samples of the stones chosen and these should represent the range of variations that are acceptable.

Types of stone

Limestones and sandstones

These stones, when used for external masonry, must be laid with the natural bed at right angles to the exposed surface. Face-bedded stones (i.e. laid with the natural bed running vertically and parallel, to the face of the wall) should not normally be used for this purpose but as joint-bedded stones they are frequently used for work to copings or for large unprotected oversailing courses or cornices. It is recommended that the suitability of the stones for this type of work be checked before selection for such purposes. Sandstone and limestone grades liable to delaminate should not be used.

Marbles
These stones tend to lose their polish when exposed to the UK climate and should not generally be used externally except in positions where they can receive regular maintenance. It is generally recommended that marbles which look well with a matt, eggshell or sand-rubbed finish should be used elsewhere.

Granites
These stones are recommended for external use. They generally retain a polish for many years in an external environment and also have a high degree of impermeability and resistance to abrasion.

Slates
This material should be assessed for durability when used in districts liable to pollution. A method of testing is given in BS 3798: 1964. When a riven finish is required considerable deviations are unavoidable. Natural markings cannot be eliminated and users should ensure that these do not imply structural weakness.

Cast stone
Cast or reconstructed stone is a form of concrete manufactured to resemble natural stone.

Dimensions

When specifying rectangular stones the 'work size' should be quoted, that is the face dimensions and thickness, to the following tolerances unless otherwise agreed with the suppliers:

1.   Units 50 mm thick or less – the length and height dimensions should not vary from those specified by more than 1.5 mm in 900 mm.
2.   Units over 50 mm thickness – the length and height dimensions should not vary from those specified by more than 3.0 mm in 900 mm.

   The total thickness of the unit should not vary from that specified by more than 3.0 mm.

Bow or twist
The face should not vary from the plane by more than 1.5 mm in 1200 mm, except that natural riven faces should not vary from the plane by more than 10 mm in 1200 mm.

Durability

The durability of stones is usually assessed based on experience and previous use. However, if doubt exists about the properties of certain stones the following tests may be carried out to obtain further information:

Group A – crushing strength $\Big\}$ Required at design stage
     wet and dry density
Group B – natural freeze thaw test – Guide to frost resistance
Group C – crystallization test – Guide to durability
Group D – porosity
     saturation coefficient $\Big\}$ Durability (e.g. Portland/Bath stone)
     microporosity

1.6 Characteristics of bricks and blocks

Fired clay units

Moisture movements
Fired clay products exhibit reversible dimensional changes dependent upon their moisture content. The wetting movement that occurs within a wall is basically controlled in a similar way to thermal movements and the actual movement will be modified by the effect of any restraints.

According to BS 5628: Part 3: 1985 the typical range of movement to be expected is generally less than 0.02 per cent within a wall, i.e. up to 0.2 mm/m which, in itself is comparatively insignificant.

Expansion
In addition to wetting movement fired clay, while cooling in the kiln, begins to take up a permanent expansion which can go on but at a greatly reduced rate, for the life of the product and this expansion is three-dimensional. The magnitude of the irreversible expansion of the unrestrained product varies with the type of clay, the method of firing the product and the maximum firing temperature. There is considerable variation in behaviour between bricks of different origins; thus, some engineering bricks, if only moderately fired, can have larger expansions than normal, i.e. up to 1.6 mm/m, though if well fired to a low absorption the same bricks will usually give very low expansions.

Recent work suggests that the ratio of brickwork expansion to brick expansion is approximately 0.6 (due to the effect of the mortar joints), provided no other source of expansion (such as sulphate expansion of the mortar) is also present. It has been found that permanent expansion starts as the units commence to hydrate during cooling in the kiln and that up to 50 per cent of the total expansion over a 2-year period can take place within a few days of the commencement of cooling. So provided that the bricks are not built into the work fresh from the kiln, moisture expansion is unlikely to present a major problem if adequate provision is made for movement in the construction, i.e. taking account of all movements which are likely to occur during the design life of the building.

It is suggested that units should not be built into the work for at least 7 days after drawing from the kiln, thus allowing the initial rapid moisture expansion to take place. However, the bricks will continue to expand and recent work at the British Ceramic Research Limited (BCRL) (Foster-Johnson, 1982; Lomax and Ford, 1983) and elsewhere predicts that the expansion at 50 years is likely to be up to 2.5 times that at 12 to 18 months. It was initially suggested by Foster and Johnson that, as a guide for permanent moisture expansion of walls of the lowest degree of restraint, three groups of brickwork might be considered, reflecting low, medium and high expansion. These figures were considered to give the greatest movements likely to be experienced during the life of a building and the designer would then be able, by experience, to reduce them progressively for increasing degrees of restraint. The figures suggested by Foster and Johnson (1982) were subsequently modified by Lomax and Ford (1988) to incorporate the results of their research and are indicated in Table 1.9.

TABLE 1.9 Moisture expansion of clay bricks

Category of brick expansion	Indicated irreversible moisture expansion
Low	<0.24 mm/m
Medium	0.24–0.48 mm/m
High	>0.48 mm/m

The figures quoted in Table 1.9 can give only a general indication of irreversible moisture expansion and it is suggested that designers request more detailed information from the manufacturers of the selected units. This information can be readily obtained if samples of kiln-fresh bricks are subjected to a steam test which has been developed by BCRL and from that test an acceptable correlation has been established between accelerated expansion and expansion measured after 5 years (see Figure 1.2).

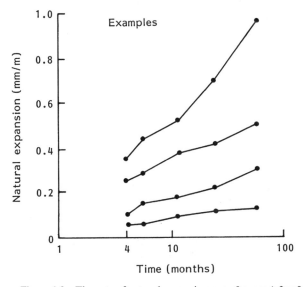

Figure 1.2   The rate of natural expansion over 5 years (after Lomax and Ford, 1988)

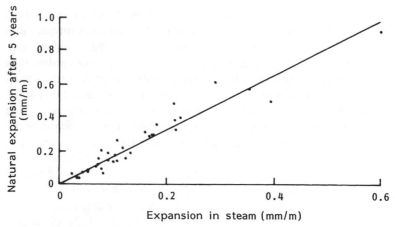

Figure 1.3 Relationship between natural expansion after 5 years and expansion in steam after 4 years (after Lomax and Ford, 1988)

The BCRL research has also indicated a linear relationship between natural expansion and the logarithm of time (Figure 1.3) and while it is perhaps unreasonable to expect an extrapolation from 5 years to, say 50 years, to be completely reliable, they consider that the test has been proven sufficiently to enable bricks to be categorized into three broad expansion categories and using the 0.6 brick/brickwork ratio, the irreversible moisture expansion of brickwork can be similarly categorized as indicated in Table 1.9.

Thermal movements
To determine the longitudinal coefficient of thermal movement of brickwork it is generally sufficiently accurate to take the same value as quoted for the units. Movement in the vertical direction may be determined by summing the values obtained by multiplying the dimensions of the bricks and the mortar by the respective coefficients indicated in Table 1.10. Other factors such as restraint, temperature ranges and orientation also need to be considered and are discussed in Chapter 4.

TABLE 1.10 Coefficient of linear thermal expansion of clay bricks and mortars

Material	*Coefficient of linear thermal movement per* °*C*
Fired clay bricks– length	4–8×10^{-6}
Fired clay bricks – height	8–12×10^{-6}
Mortars	11–13×10^{-6}

Concrete and calcium silicate bricks and blocks

Moisture movements
Concrete and calcium silicate bricks and blocks shrink after manufacture due to drying shrinkage and carbonation. Some of the shrinkage is reversible in that units of this type will expand if they become wet but this wetting expansion rarely causes difficulty because it is less than the initial drying shrinkage. Concrete units which are not autoclaved are subject to a slow, non-reversible carbonation shrinkage

caused by chemical reaction of carbon dioxide from the atmosphere with some of the products of hydration of the cement. Drying shrinkage should be taken into account when making provision for movement unless the units are intended to remain continuously wet. It is normally recommended that concrete bricks and blocks should be stored before use, exposed to the wind but protected from moisture, for at least 4 weeks at normal temperature and longer in cold weather; slightly shorter periods of storage may be adequate if the units have been cured in low pressure steam and cooled so that they have an opportunity of drying. Autoclaved concrete and calcium silicate products need only sufficient storage to allow them to cool.

BS 5628: Part 3: 1985 quotes the figures given in Table 1.11 for moisture movement of concrete and calcium silicate masonry units. The extent of carbonation and the subsequent movement depends on the permeability of the concrete or the calcium silicate and on the ambient relative humidity. The magnitude of this movement in dense concrete masonry units and in autoclaved concrete or calcium silicate units is extremely small and may generally be neglected.

TABLE 1.11 Shrinkage of concrete and calcium silicate units

Material	Shrinkage percentage of original (dry) length[+]
Autoclaved aerated concrete masonry units	0.04–0.09
Other concrete masonry units	0.02–0.06
Calcium silicate bricks	0.01–0.04

Note: These figures are obtained from tests carried out as described in BS 1881: Part 5: 1970.
[+]The higher figures are the limits specified in the appropriate BS for quality control purposes and should not be taken to represent the movement of units in a wall.

In open textured masonry units and mortar shrinkage due to carbonation may be between 20 per cent and 30 per cent of the initial free moisture movement.

Thermal movements
The coefficients of linear expansion quoted in Table 1.12 relate to an unrestrained condition and the imposition of restraint, whether from within the wall by friction or external restraint, will modify considerably any movement based on these data. It is important to recognize that concrete and calcium silicate units vary dependent on the type of material used, the method of manufacture and the mix proportions.

To determine the longitudinal coefficient of thermal movement of masonry it is sufficiently accurate to take the same value as quoted for the units. Movement in the vertical direction may be determined by summing the values obtained by multiplying the dimensions of the masonry units and the mortar by the respective coefficients.

TABLE 1.12 Coefficients of linear thermal expansion – concrete and calcium silicate units

Material	Coefficient of linear thermal movement per °C
Concrete masonry units	$7{-}14 \times 10^{-6}$
Calcium silicate masonry units	$11{-}15 \times 10^{-6}$
Mortars	$11{-}13 \times 10^{-6}$

1.7   Specific problems

Fired clay bricks

Efflorescence
Efflorescence or deposits of soluble salts may form on the surface of certain types of fired clay bricks. These salts usually show as loose white powder or as feathering crystals. Occasionally they appear as a hard glossy deposit covering and penetrating the brick faces.

Bricks may be tested for liability to efflorescence quite simply using the method laid down in BS 3921: 1985. In using this test it must be realized that, although it may readily indicate certain degrees of liability to efflorescence, in the laboratory, this may not be a reliable means of determining what will occur in practice.

There are undoubtedly numerous reasons for this phenomenon but perhaps the main one is that although evaporation from faces of the brick other than that which will appear as the exposed face in the work, is prevented by surrounding them with an impermeable sheet, such as polythene. This method of test does not necessarily simulate site conditions, where adjoining surfaces undoubtedly have a considerable affect.

The subject of efflorescence on brickwork is discussed in greater detail in Chapter 6.

Soluble salts
Limits on the soluble salts content of bricks may be necessary when walls constructed of the units are to be sited in exposed conditions, thus minimizing the risks of sulphate expansion of Portland cement mortar and efflorescence on the brickwork. Although the reasons for these forms of distress are understood, the limits laid down in BS 3921: 1985 for designation L bricks (i.e. bricks with low soluble salt content) may or may not be suitable. In some circumstances it would appear that bricks with a total soluble sulphate content of well under 1 per cent have given severe trouble in sulphate expansion of mortars and/or renders whereas in others, bricks with soluble salt contents of as much as three times this amount have been used without arousing comment.

Similar evidence has been noted on particular salts, e.g. potassium sulphate. For instance, there has been complete absence of complaints over an extended period when bricks containing 0.25 per cent soluble potassium have been used. Elsewhere trouble has arisen with bricks containing less than 0.25 per cent.

Sulphates of sodium or magnesium are more troublesome than those of calcium or potassium.

The presence and quantity of soluble sulphates depends very much on the source of the raw material and the degree of firing in the kiln. Underfired clay bricks are more likely to contain an excess of soluble salts than well-fired units.

The subject of sulphate attack on mortars and renders is considered in more detail in Chapter 6.

Liability to frost attack
As stated earlier three designations of frost resistance are specified in BS 3921: 1985: F – frost resistant, M – moderately frost resistant, and O – not frost resistant.

Considerable research has been carried out on frost resistance of fired clay products but no individual test is currently included in the BS 3921: 1985. According to Stupart (1989) the critical properties of the bricks which need to be considered are:

Pore size distribution and porosity
Moisture content (degree of saturation)
Elasticity, E value
Tensile and compressive strength
Degree of firing
Presence of flaws such as laminations

There is considerable interdependence of many of these factors. As long ago as 1929, McBurney carried out an extensive series of tests at the US Bureau of Standards on water absorption saturation coefficient and crushing strength of samples. He found that, if the samples were arranged in order of saturation coefficient, those with the highest saturation coefficient were of poor, those with the lowest, good durability. It was not possible however to draw a line at any particular point without either excluding some satisfactory units on the one hand and unsatisfactory ones on the other. The same was true of water absorption and crushing strength. Artificial freezing and thawing tests have also been tried and a reasonable close correlation established between the performance in the laboratory and under actual site conditions.

The best evidence of ability to withstand frost damage is provided by 'brickwork' which has been in service for some years under the same or similar site conditions as those proposed by the designer of a building and this evidence can usually be provided by the brickmaker or supplier.

When it is not possible to provide evidence of frost resistance by the above method a 'freeze–thaw test' carried out by the BCRL may be helpful or the old, but sometimes conservative, method of selecting bricks having a crushing strength of not less than 48.5 N/mm^2 and/or a unit with a water absorption not greater than 7 per cent should provide a frost-resistant unit.

Drier and kiln scum
This phenomenon is rarely seen on other than common bricks but was a frequent problem in years gone by, particularly on bricks made from clays of marine origin. When the bricks are made from such clays the salts dissolve in the mixing water and can be deposited on the surface when the bricks dry in the form of a white or pale-coloured coating known as scum. The deposit of salts so formed combines with the clay at the surface of the brick during burning so that it becomes insoluble and cannot be washed off. The deposit appears only during drying, so that the areas where bricks are in contact in the drier or kiln are free from scum because the mixing water with its dissolved salts cannot escape through them creating what is known as kiss marks.

Kiln scum is closely related to drier scum and may form if the green bricks are exposed in the early stages of burning to gasses containing sulphur compounds derived from sulphur when coal firing is used. If the bricks are not sufficiently warmed by hot clean air from the cooling zone of the kiln before the chamber is taken on to the firing cycle moisture may condense on the cool bricks and dissolve the acidic sulphur oxides, forming an acid solution that reacts with the clay and so forms a scum.

Dunting
After a brick has been fired it has to be cooled to an ambient temperature so that it can be handled and is ready for use. During the cooling process bricks can sometimes crack as a result of too rapid a reduction in temperature. This form of cracking is known as dunting and if built into a wall can often be mistaken by the unwary as a form of structural distress.

Underfired bricks
It is generally agreed that the most important stage in the manufacture of fired clay bricks is the burning or firing process. If the brick is underfired so that it is soft, muddy brown or bright salmon pink in colour and gives only a dull sound when struck, it is likely to have poor durability and may contain a higher soluble salt content with all the attendant potential problems. However, it should be noted that some soft mud type bricks do not have the same ring when struck and although initially soft may have excellent durability. It should also be borne in mind that a brick might be eminently satisfactory in one part of the country or even one location within the external envelope of a building whereas it may rapidly fail when exposed in another location.

Black heart bricks
If during the firing process the temperature rises too rapidly the brick starts to vitrify and the surface pores close during the oxidation stage before the various gaseous oxides have diffused out. Such a brick can then be recognized by its having a black core (or black heart). The colour may be partly due to unoxidized carbon, as well as the condition of the iron in the clay. When oxygen is deficient, much of the iron will be present in the form of dark bluish ferrous compounds and sometimes the iron is deprived of its oxygen altogether and reduced to the form of a metal. If the brick is soaked in water, rust staining may occur on the face of the brick.

Black-cored bricks are not necessarily less serviceable than similar bricks which have been fired more slowly and are free from black cores but such bricks are more liable to soften rapidly at high temperatures and there is an increased tendency to bloating, i.e. swelling up as a result of the formation of gas bubbles in the body. Such bricks are, of course, unacceptable and are generally, but not always, removed during the works selection process.

Some black-cored bricks contain an unacceptable level of water soluble sulphates with potential attendant problems of persistent efflorescence and the danger of sulphate attack of mortars and renders which contain Portland cement. Designers would be well advised to seek assurances from brick manufacturers supplying bricks with black cores that the product has acceptable soluble sulphate levels and a good record of durability.

Some black-cored bricks exhibit colourful bruise-type blemishes on the exposed faces which may or may not be acceptable, as well as black specks which may expand and cause the face of the brick to spall in a similar manner to lime blowing.

Cracking/crazing of brick faces
Condensation on freshly set green bricks (i.e. unfired bricks) during the drying process can, under certain conditions, be absorbed into the exposed brick faces and subsequent drying may produce cracking or surface crazing. In some cases the bricks may soften to such an extent that the green bricks deform under their own weight.

Crazing of extruded bricks can also occur if oil from the balancing die adheres to the brick faces during the extrusion process.

Laminations which give rise to flaking, cracking or fractures appear mainly in extruded machine-made bricks. An inherent feature of all blades, or screw, extruders is that the clay enters the mouthpiece in the form of a helical coil and weaknesses sometimes occur due to uneven packing of the clay at the brick corners. This makes the bricks liable to delamination in frost and S-type cracking sometimes also occurs on the bed faces due to the same manufacturing defect. Cracking of bricks can also occur due to defective firing and/or during the cooling process.

Ragged or crumbly edges and corners
Stiff-plastic clay bricks may have ragged or crumbly edges and corners if the clot mould (clay mould) is not completely filled. This is because there will be insufficient clay in the clot to form a good brick. This defect is usually due to a worn auger or wing knife.

Similarly, uneven distribution of clay in the clot mould will, on pressing, lead to variations in density in various parts of the brick.

Chamfered arrises
A chamfered arris sometimes occurs during the manufacturing process with soft mud bricks and is caused by particles of unfired clay sticking to the mould. Many bricks of this type have a slight chamfer to the arris and this is accepted as a normal characteristic. However, if the chamfering is pronounced the brick would be rejected or sold as a 'second'. This characteristic is usually confined to one face and if considered unacceptable the bricklayer usually lays the imperfect face towards the inner leaf in cavity wall construction.

Lipping on arrises
Lipping may be caused during the manufacture of soft mud bricks. This feature sometimes occurs adjacent to the bed face without a frog and is sometimes seen on the header face due to a sand deposit causing the unfired clay to be dragged over the side of the mould after being pressed. This results in a lip being formed which is usually not very noticeable. Similarly, if the unfired clay is a little on the dry side and the mould box is tight under the feeder tower, some tearing of the clay may occur but once again the resulting lipping tends to occur more on the header face than the stretcher face. This feature is not usually very noticeable and is again generally accepted as a characteristic of this type of brick.

Thumbing

Once again this is a minor defect which sometimes occurs during the manufacture of soft mud bricks on some header faces but according to the manufacturers never on both ends. After the automatic washing of the mould boxes occasionally a globule of water adheres to one end of the box. When sprayed with sand the water adsorbs a lump of sand which subsequently displaces the plastic clay prior to firing. In normal cavity wall construction only one header face is ever exposed; therefore providing the bricklayer lays the corner bricks with the unaffected face exposed there is no reason to reject any bricks displaying this characteristic.

Holes in faces

This is due to a characteristic of the machine used for the soft mud process due to a similar phenomenon to that described for thumbing. The holes are normally an acceptable feature of this type of brick and if pronounced are only a problem if they appear on two opposite faces. Normally the bricklayer selects the unaffected surface for the external face of the wall.

Characteristics of clamp-fired Stock bricks

Stock bricks are now made using a modern version of the centuries old method. The clay for the brick is mixed with a small quantity of fuel (usually coke dust) and water added to produce a soft mud. The mud is then poured into a sanded mould. After moulding the bricks are oven dried and the green bricks subsequently fired in what are known as clamps. The clamps consist of a large stack of the unfired bricks with several horizontal layers of fuel interspersed within the mass. One end of the base fuel is ignited and the fuel and bricks burn, forming sufficient heat to fire the bricks.

This process does not produce bricks of uniform colour or texture as in the more controlled tunnel and similar kilns. Because of the method of firing there are dark marks and small holes in the surface where the fuel, which has been incorporated into the clay, has burnt out. It is not uncommon for some distortion of the bricks to occur or even for several units to be formed into a mass due to the firing process and the weight of the overburden. After firing the bricks are sorted into colour and quality groupings.

Each manufacturer has his own method of assessing quality but 'seconds' may have chipped edges, be slightly irregular in size, have an uneven texture and variable colour. First quality bricks will generally have four acceptable faces and 'rejects' will be more variable than 'seconds', contain damaged faces and bricks of irregular size.

Even 'first quality' clamp fired 'Stock bricks' may not comply with all the requirements of BS 3921: 1985.

Pebbles and stones

Some Boulder clays contain pebbles and/or stones and provided that these aggregates do not appear on or near the surface of the faces to be exposed in the walling the bricks are generally considered acceptable. If on the other hand the pebbles or stones are near the brick faces, firing cracking may occur or if the aggregates are exposed the bricks may not be acceptable from an aesthetic point of view and may be susceptible to frost damage.

Exposed pebbles or stones and/or firing cracking on the bed faces of the bricks do not usually have a deleterious effect provided the defect is not adjacent to the brick face. This defect is rarely seen in modern fired clay bricks.

Lime blowing
If lime is present in the green bricks in nodules, e.g. as limestone pebbles or fossils, when the bricks are fired in the kiln, the carbonate burns to quicklime (calcium oxide) and if the lumps of lime are allowed to slake slowly by absorbing moisture from the air they are liable to expand and cause splitting or spalling of the fired bricks.

This defect can be overcome at the brickworks, provided that the amount of lime is not too great, by grinding finely and/or by immersing the fired bricks in water (docking), so that the lime is slaked to a plastic mass that cannot exert sufficient pressure on the brick to cause damage.

Stains
Vanadium staining. Some clays used for brickmaking contain traces of vanadium minerals, which when dissolved, produce a green stain on the bricks. This staining is rare but when it leaches out onto the faces of light-coloured bricks tends to resemble moss and is frequently mistaken for an algae growth.

Iron staining. Brown rust-like streaks can occur on certain types of kiln-flashed purplish, antique or multi-colour bricks as a result of the reducing conditions employed during the firing process to attain the desired colouring effects. The rust-like stains, which show up on the mortar joints, occur due to precipitation of ferric compounds from the soluble ferrous solutions (formed by acid rain water dissolving ferrous compounds in the brick) due to the action of lime in the mortar. These stains should not be confused with those emanating from contaminated aggregate in some mortars.

Manganese staining. This type of staining of clay bricks is similar to iron staining but is generally dark brown or black in colour.

Concrete bricks and blocks

Shrinkage
Concrete bricks and blocks shrink after manufacture due to drying shrinkage and carbonation (a chemical reaction between carbon dioxide from the atmosphere and some products of hydration of the cement). Drying shrinkage should be taken into account when making provision for movement unless the bricks or blocks are to remain continuously wet (e.g. below d.p.c. level).

Concrete bricks and blocks should be covered in transit (preferably by wrapping) and on site and stored before use. Bricks should be stored for at least 7 days and blocks for at least 4 weeks, both types of unit requiring a longer storage period in cold weather; specially cured units and autoclaved, aerated concrete blocks may require less storage time.

If wet units are used or moisture is absorbed during the construction process, the walls should be allowed to dry before plastering or dry lining to reduce the danger of shrinkage damage.

It is important that the drying shrinkage of concrete masonry units does not exceed the limits given in BS 6073: Part 1: 1981.

Frost resistance
Pre-cast concrete units in general possess good frost resistance provided they are selected in accordance with the recommendations of BS 5628 : Part 3: 1985.

The resistance of concrete units to frost depends on their various properties but the main factors are the degree of saturation and the pore structure of the cement paste. As the temperature of saturated hardened concrete decreases the water held in the capillary pores in the cement paste freezes in a similar manner to the freezing in the capillaries in rock, and expansion of the concrete surface takes place. Repeated freeze–thaw cycles have a cumulative effect that results in spalling of the faces.

Acid rain
Acid rain and other acid fumes present in the atmosphere can affect the exposed surfaces of concrete units by dissolving and removing part of the set cement leaving behind a soft and mushy mass.

Sulphate attack
Dense concrete units are not usually subject to sulphate attack unless buried in ground with an exceptionally high sulphate content. Some less dense units may be vulnerable and if in doubt the manufacturer of the units should be consulted.

Sea water contains sulphates which may have a deleterious effect and may also be responsible for efflorescence on the units.

Surface colour
Some concrete units have an applied surface colour which contrasts with the main body of the units and in aggressive environments erosion of the face can sometimes occur producing colour variation over the wall surface.

Surface damage
Surface damage, in addition to causing unsightly arrises, can expose the aggregate in dense units finished with an applied finish.

Efflorescence
Efflorescence occurs on some concrete units in a similar manner to clay bricks but usually, with light or cement coloured units, the problem is less apparent than with the darker coloured units.

Calcium silicate bricks

General
The physical properties of calcium silicate bricks are largely characterized by the weight, compressive strength, resistance to frost and the definition of the arrises.

These properties are determined by the raw materials and the processes used in manufacture, including preparation processes, moulding and hardening. In particular, the brick weight, compressive strength and frost resistance are largely determined by the nature of the moulding process.

Frost resistance
The frost resistance of Class 2 bricks is normally considered adequate for normal walling and that of bricks of Classes 3 to 7 for the most severe conditions of exposure. In conditions of severe frost, calcium silicate bricks which have become saturated with a strong solution of a chloride are liable to disintegration.

A somewhat different effect occurs with units that have been repeatedly wetted with sea water or spray. The gradual accumulation and repeated recrystallization of salts over the years is found to cause a gradual erosion of the surface.

Sulphate attack
Only exceptionally when subjected to high concentrations of magnesium sulphate, are calcium silicate bricks liable to attack by sulphates occurring naturally in soils. In heavily polluted atmospheres, calcium silicate bricks can, in sheltered positions, gradually develop a skin containing calcium sulphate. When this occurs erosion or blistering and flaking of this skin can take place in the same manner as limestone units over a long period of time under adverse conditions.

Colour
Calcium silicate bricks can be manufactured in a wide range of through colours, i.e. the bricks are homogeneous, and exhibit practically the same colours when cut or chipped, as the pressed surfaces.

Shrinkage
The drying shrinkage of calcium silicate bricks is generally less than that of plain dense concrete units and much less than that of lightweight concrete units.

Efflorescence
Efflorescence is rarely seen on calcium silicate bricks.

1.8   Method of specifying the units

To ensure that the designer's requirements are met and to avoid disputes over the quality and properties of the units when incorporated in the works the following particulars should be clearly stated in the contract documentation:

Clay bricks

(1)   Specify that the bricks are to comply with the appropriate requirements of BS 3921: 1985 *Clay bricks*. Marking BS 3921: 1985 on or in relation to a product is a claim by the manufacturer that the product has been manufactured in accordance with the requirements of the standard.
(2)   State the quantity (making allowance for breakage) and handling requirements, such as palletization, strapping or mechanical off-loading.

(3)   State the name, trade mark or other means of identification of the manufacturer.

(4)   State the type of brick, i.e. solid, cellular, frogged or perforated.

(5)   State the variety, i.e. Engineering A or B, d.p.c. brick or facing, handmade or common.

(6)   State the compressive strength if it is required to be greater than 5 N/mm^2.

(7)   State the name of the brick, e.g. Red Multi.

(8)   State the durability designation, e.g. FL, FN, ML, MN, OL or ON.

(9)   If special shapes (e.g. splayed quoins or closure bricks) are required it may be necessary to agree the shape with the manufacturer.

If standard special bricks are required the appropriate reference in BS 4729: 1971 should be quoted.

Calcium silicate bricks

(1)   Specify that the bricks are to comply with all the requirements of BS 187: 1978 *Calcium silicate (sandlime and flintlime) bricks.*

(2)   State the quantity (making allowance for breakage) and handling requirements, such as palletization, strapping or mechanical off-loading.

(3)   State the brick size required and their designation (i.e. loadbearing or facing).

(4)   State the compressive strength or strength class required, if no strength is specified the lowest strength is implied. Also, if special means of identifying the bricks for strength is necessary the colour code shown in Table 1.6 should be requested.

(5)   State the type of brick, e.g. completely solid, with a frog in one bed face, etc.

(6)   State the type of material, e.g. sandlime or flintlime. Also if required, the colour or pigment to be used.

(7)   If special shapes (e.g. splayed quoins or closure bricks) are required it may be necessary to agree the shape with the manufacturer.

(8)   When the purchaser requires bricks having special performance properties these should be specified.

Concrete bricks and blocks

(1)   Specify that the concrete bricks or blocks comply with all the appropriate requirements of BS 6073: Part 1: 1981 *Specification for pre-cast concrete masonry units.*

(2)   State the quantity (making allowance for breakage) and handling requirements, such as palletization, strapping or mechanical off-loading.

(3)   State the work size required.

(4)   State the compressive strength required. For concrete blocks where this is greater than the minimum quoted in BS 6073: Part 1, i.e. 2.8 N/mm^2. In any case, for blocks of thickness 75 mm or greater the purchaser is required to specify the minimum compressive strength required. For example, where a purchaser specifies that a block shall be of a minimum compressive strength of 7.0 N/mm^2 this means that a sample of 10 blocks tested in accordance with BS 6073: Part 1: 1981 shall have an average crushing strength of 7.0 N/mm^2 and the corresponding lowest crushing

strength of any individual block shall be not less than 80 per cent of 7.0 N/mm^2, i.e. 5.6 N/mm^2.

(5)   For bricks state the type, e.g. completely solid, solid with a frog in one bed face, etc. For blocks the type will be solid, cellular or hollow.

(6)   State the type of material, e.g. for bricks, dense concrete, reconstructed stone, etc. Also if required, a restriction on the materials and pigments to be used. For blocks, dense, lightweight (lightweight aggregate or autoclaved aerated concrete) or reconstructed stone.

(7)   If special shapes (e.g. splayed quoins or closure bricks) or bricks for which special tolerances are necessary, such as exposed aggregate for fair faced masonry, it may be necessary to agree the shape or tolerances with the manufacturer.

(8)   When the purchaser requires bricks or blocks having special performance properties these should be specified. Also if special means of identifying the units is necessary.

References

BCRL Special Publication No 56. *Model specification for clay and calcium silicate structural brickwork.* January 1988.

BS 3921: 1985 *Clay bricks.* BSI, London.

BS 6073: Part 1: 1981 *Pre-cast concrete masonry units. Specification for pre-cast concrete masonry units.* BSI, London.

BS 5628: Part 3: 1985 *Use of masonry – materials and components design and workmanship.* BSI, London.

BS 187: 1978 *Specification for calcium silicate (sandlime and flintlime) bricks.* BSI, London.

BS 5628: Part 1: 1978 *Structural use of masonry – unreinforced masonry.* BSI, London.

BS 3978 : 1964 *Coping units (of clayware, unreinforced cast concrete, unreinforced cast stone, natural stone and slate.* BSI, London.

BS 1881: Part 5: 1970 *Methods of testing hardened concrete for other than strength.* BSI, London.

BS 4729: 1971 *Shapes and dimensions of special bricks.* BSI, London.

Foster, D. and Johnson, G. D. (1982) Design for movement in clay brickwork in the UK. *Proceedings British Ceramic Society*, No 3, September.

Lomax, J. and Ford, R. W. (1983) Investigations into a method of assessing the long term moisture expansion of clay bricks. *Transactions and Journal of the British Ceramic Society.* Vol. 82.

Lomax, J. and Ford, R. W. (1988) A method for assessing the long term moisture expansion characteristics of clay bricks. *Proceedings of the 8th International Brickblock Masonry Conference*, Vol. 1, September.

McBurney, J. W. (1929) *Weathering quality of bricks.* US Bureau of Standards.

Stupart, A. W. (1989) A survey of literature relating to frost damage in bricks. *Journal of the British Masonry Society,* Vol. 3, No. 2, 42–84.

2
Mortars

The selection and correct specification of the constituents of mortars play an important part in the subsequent performance of masonry walls and time spent in considering the materials to be used is never wasted.

2.1 Cements

In modern masonry construction the cement used is generally either ordinary or rapid-hardening Portland cement to BS 12: 1978, or Portland blast-furnace cement to BS 146: Part 2: 1978, providing the slag content does not exceed 35 per cent. (British Ceramic Research, 1988).

Special cements have particular advantages (and disadvantages) such as particular rates at which they set and harden; other types resist attack by some chemicals and there are cements which offer a choice of workability or colour. The normal rules of good building practice apply when using special cements and none of them will prevent or cure defects caused by bad workmanship or failure to make the mortar correctly.

The following types of cement may be used according to BS 5628: Part 3: 1985.

Portland cement (ordinary and rapid hardening to BS 12: 1978).
Portland blast-furnace cement to BS 146: Part 2: 1978.
Sulphate-resisting Portland cement to BS 4057: 1980.
Masonry cement to BS 5224: 1976.

Ordinary Portland cement

A typical analysis of ordinary Portland cement would indicate the presence of four principal compounds: lime (CaO) 60–70 per cent; silica (SiO_2) 20-25 per cent; alumina (Al_2O_3) 3–8 per cent; and iron oxide (Fe_2O_3) 2–3 per cent. Other materials such as magnesia, sulphuric anhydride and alkalis, etc. may be present in the raw materials used for cement manufacture, but the quantities in the final product are limited by the British Standard for Portland cement BS12: 1978. The presence of more than

26

4 per cent magnesia (by weight) may lead to unsoundness of the resulting mortar. Ordinary Portland cement is the least expensive and by far the most widely used type of cement and is suitable for all normal purposes.

Rapid-hardening Portland cement

This cement is very similar in manufacture and composition to ordinary Portland cement but has a higher tricalcium silicate (C_3S) content and cement clinker is usually more finely ground. The British Standard BS12: 1978 specifies a higher specific surface area for this cement than for ordinary Portland cement, but as a general rule a high fineness is encountered. The name rapid hardening is somewhat misleading and it would perhaps be more aptly described as high early strength cement.

Strength developed at an age of 3 days is of the same order as the ordinary Portland cement 7-day strength, subject to the water/cement ratio being constant. Both cements have approximately the same ultimate strength.

Extra-rapid-hardening Portland cement

Extra-rapid-hardening Portland cement is intended for cold weather concreting only and is not suitable for mortars. The cement is produced by intergrinding up to 2 per cent of calcium chloride with rapid-hardening Portland cement and must therefore be stored under dry conditions, preferably being used within 28 days of dispatch from the cement works. Calcium chloride is deliquescent and must not be used in mortar.

Low heat Portland cement

Low heat Portland cement BS 1370: 1979 is intended for structures where a large concrete mass is necessary and for this reason it is desirable to limit the rate of heat evolution. It is not generally suitable for mortar as the rather lower content of the more rapidly hydrating compounds, C_3S and C_3A, results in a slower development of strength of low heat cement as compared with ordinary Portland cement, but the ultimate strength is unaffected.

White Portland cement

White Portland cement is generally used in mortars for aesthetic reasons where a white or pastel colour is required in the mortar joint. Coloured cement usually consists of white Portland cement interground with between 2 per cent and 10 per cent pigment, the addition of such non-cementitious material can affect the properties of the mortar.

White Portland cement is more expensive than ordinary Portland cement as contamination of the cement with iron during grinding must be avoided and for this reason more costly grinding techniques must be adopted.

Portland–pozzolana cement

Portland–pozzolana cements are manufactured by intergrinding or blending mixtures of Portland cement and pozzolana, the pozzolanic material containing silica in a reactive form. Materials commonly used as pozzolana are: volcanic ash, pumicite, opaline shales and cherts, calcined diatomaceous earth, fired clay, fly ash, etc. The rate of strength development depends on the activity of the pozzolanas and on the percentage of Portland cement in the mixture.

Portland blast-furnace cement

Portland blast-furnace cement is not produced in England, but it is produced extensively in Scotland. It is made by intergrinding Portland cement clinker and blast-furnace slag. BS 146: Part 2: 1978 limits the proportion of granulated blast-furnace slag to 65 per cent of the weight of the mixture. British Standard requirements for ordinary and blast-furnace Portland cements are the same but the latter tends to be finer. Even so, the rate of hardening of Portland blast-furnace cement tends to be somewhat slower during the first 28 days. At later stages there is little difference between the strength of the two cements.

Another cement sometimes classified under the same heading consists of blast-furnace slag and lime ground together; the term Portland should never be applied in this case.

Sulphate-resisting Portland cement

Sulphate-resisting Portland cement is similar in manufacture and composition to ordinary Portland cement, the only special requirements of the British Standard BS 4027: 1980 being limits placed on the tricalcium aluminate (C_3A) content and on the specific surface area of the cement. The reason for the limitation on the C_3A content is that this substance can react with sulphates in solution to form the compound calcium sulpho-aluminate or 'Ettringite'. The reaction is accompanied by an increase in volume which is sufficient, usually, to cause gradual disintegration of the mortar. Some ordinary Portland cements have a low C_3A content and offer excellent resistance to sulphate attack but manufacturers will not name the source of such cements for obvious reasons.

Masonry cement

Masonry cement is defined by the International Organisation for Standardisation as 'A finely ground mixture of Portland cement or other appropriate cement and of materials which may or may not have hydraulic or pozzolanic properties, and which may or may not include air-entraining agents, plasticizers, water repellent substances, etc. It is characterised by certain physical properties such as slow hardening, high workability and high water retentivity, which makes it especially suitable for masonry work'.

Masonry cements marketed in the United Kingdom contain 75 per cent ordinary Portland cement, 25 per cent of a fine filler such as ground limestone and an air-entraining agent. British Masonry cements contain a higher proportion of ordinary Portland cement than those of many other countries and for this reason it is unnecessary to gauge additional Portland cement to obtain mortar of high strength. The strength of masonry cement mortars in the UK is in fact, only controlled by altering the proportion of sand, the normal range of volume proportions being from $1:2\frac{1}{2}$ to 1:7 masonry cement:sand. Masonry cement is intended chiefly for use in mortars for brick, stone and concrete block construction. When masonry cement is mixed with sand, it produces a smooth, plastic and cohesive mortar characterized by a lower rate of strength development than that of ordinary Portland cement. It may not, therefore, be suitable for all forms of structural masonry or for special purpose mortars where high strength is required. One specification (British Ceramic Research Limited, 1988) limits the use of masonry cement mortars for structural brickwork to 1:4½ masonry cement:sand, i.e. the equivalent of a 1:1:6 Portland cement:lime:sand [mortar designation (iii)]. The reason for the limitation is that experience indicates little variation in strength up to the equivalent of a mortar designation (iii) but for stronger mixes the difference becomes more pronounced.

For non-structural masonry, masonry cement mortars have the advantage that the lower strength development accommodates movements due to shrinkage, solar radiation, moisture expansion and minor amounts of settlement in the mortar joints, thus preventing unsightly cracking.

Water-repellent Portland cement

This cement is sometimes used in rendering to reduce water penetration. However, effective damp-proof courses and flashings are still necessary when these cements are used.

Mix proportions and mixing time must be carefully controlled as these cements contain an air-entraining agent. Over mixing will result in a higher air content and may result in a lower mortar strength.

2.2   Limes

The manufacture and use of lime for mortar and plaster is an ancient art. Lime was used by the Ancient Egyptians for plaster, some of which has been preserved until today. The Romans used lime extensively for mortar and plaster and for making concrete. Some of their great buildings are evidence of its durability. In the fifteenth century a renaissance in the use of plastering began and during succeeding centuries decorative internal plasterwork and external stucco were developed to a great art.

Today, lime is used for mortar, plastering and external rendering although generally in combination with Portland cement as the principal binding agent. The combination of lime and Portland cement or lime and gypsum plaster has extended the usefulness of these materials to give, in many cases, better results than any of them used separately.

Lime confers on mortar and plaster mixes in which it is incorporated the essential properties of good plasticity and workability, which give ease of application with good adhesion to masonry units. A property associated with good workability is that of good water retentivity, a property essential in mortar when used in conjunction with 'high suction' bricks or blocks. The mortar or plaster mix in which it is incorporated will not readily lose its mixing water, so that ample time is provided for working before the mix becomes stiff. Consequently the risk of the cement or gypsum plaster becoming starved of water before it has had time to become hydrated and set is negligible, unlike straight cement mortar and or plaster mixes on high suction backgrounds which can be starved of moisture prior to setting.

Adding lime to a cement:sand mix makes it more workable, water-retentive and less liable to crack. Its addition to gypsum gives similar improvements in workability and water retentivity coupled with its ability to reduce expansion and act as a useful barrier to efflorescence.

Another distinctive feature of lime is its high volume yield, i.e. the unit weight of lime gives a much larger volume of plaster material than can be obtained by many other cementitious materials.

Source and composition of materials

The raw material from which building lime is manufactured is found either in the form of chalk or limestone. The main chemical constituent of naturally occurring chalk is calcium carbonate which may form 85 per cent or more of the material. The highest calcium carbonate contents are found in white chalk; grey chalk (or grey-stone) has a lower calcium carbonate content together with some clay-like material.

Making quicklime and hydrated lime

Limestone or chalk is burnt in a kiln, during which process the high temperatures decompose the calcium carbonate, carbon dioxide is driven off as a gas, and the material known chemically as calcium oxide or commercially as quicklime remains, i.e.:

$$CaCO_3 \quad = \quad CaO \quad + \quad CO_2$$

Calcium   Calcium   Carbon
Carbonate   Oxide   Dioxide

Limestones and chalks vary a good deal in their make-up; although basically consisting of calcium carbonate, they all contain other ingredients in varying amounts, which may change the characteristics of quicklime and hydrated lime made from them.

Lime is supplied in two principle forms:

(a)   quicklime
(b)   hydrated lime

Quicklime is sold in lump and ground form. When the quicklime is drawn from the kiln, the most suitable lumps are marketed as lump quicklime without further treatment.

It is also sold in bagged powder form and marketed as ground quicklime – this should not be confused with hydrated lime powder. Hydrated lime is produced from quicklime by causing it to react with water, the amount of which is so controlled that it is sufficient to slake the quicklime completely, while at the same time producing a fine powder, i.e.:

$$CaO \quad + \quad H_2O \quad = \quad Ca(OH)_2$$

Quicklime   Water   Calcium hydroxide or
                                  hydrated lime

The process takes place in a machine known as a hydrator and the end product is subsequently refined to eliminate any coarse and harmful constituents prior to being bagged and sold as hydrated lime powder.

Hydraulic properties
Both quicklime and hydrated lime may be either non-hydraulic or semi-hydraulic for the following reasons. A pure chalk when burnt gives a quicklime containing practically no impurities, whereas other chalks, e.g. grey chalk, may contain a proportion of clay-like materials which, during burning, combine with a small proportion of the lime and are converted into compounds similar to those present in Portland cement. Naturally occurring chalks have constituents which vary between these two extremes.

Lime produced from the pure varieties of chalk is therefore non-hydraulic whereas that from the impure varieties is hydraulic. The degree of hydraulicity depends on the proportion of clay-like materials present in the chalk and also the burning conditions.

BS 890: 1972 includes tests which determine the classification of building limes dependent upon their degree of hydraulicity. If a non-hydraulic lime and sand specimen is kept under damp conditions it will not harden, whereas if a similar specimen is made from a semi-hydraulic lime and kept under identical conditions it will harden slowly but never becomes very hard like a cement:lime:sand mortar.

Storage

Hydrated lime, especially in paper sacks, should be stored in cool, dry, draught-free conditions for the minimum of time (certainly not more than a few weeks) since atmospheric carbon dioxide can combine with the lime and reduce its effective properties. If external storage cannot be avoided, the bags should be placed on a raised platform and covered with weatherproof covers weighted and tied down.

Quicklime putty

Research has shown that mortars containing a proportion of quicklime putty have superior workability to those containing dry hydrated lime or even pre-soaked hydrated lime. However, the preparation of quicklime putty does take time and is consequently not very popular on many building sites.

The most reliable method is as follows:

(1)   The slaking vessel should be filled with water to a depth of approximately 300 mm and should be large enough to permit free stirring of the mix.
(2)   Enough quicklime should then be added to cover the bottom of the vessel and to come about halfway to the surface of the water. It is important that the lime is added to the water and not the water to the lime.
(3)   The mixture should be immediately thoroughly stirred, care being taken that no lime becomes exposed above the surface of the water. If such exposure seems likely to occur, or if the escape of steam becomes too violent, more water, of which a plentiful supply should always be to hand, should be immediately added.

(4) Water and lime should be added alternately with continuous stirring until the required quantity of milk of lime is produced.

(5) When all reaction has ceased, stirring should be continued for at least 5 minutes. The resulting milk of lime should be run through a 3-mm square mesh sieve into a suitable storage bin and must be allowed to mature undisturbed for a period of at least 2 weeks. It will fatten up to lime putty and should be protected from drying out. Quicklime is especially dangerous and special precautions should be taken when handling this material such as the use of safety goggles, face masks and gloves, etc.

Hydrated lime

Hydrated lime should be used:

(1) As a putty. The hydrated lime should be added to water (not the water to the lime) in a clean container, thoroughly mixed to the consistency of thick cream and allowed to stand not less than 16 hours. Any excess water will rise to the top of the mix and should be siphoned or decanted off before use.

(2) Mixed dry with sand and then water added. The material produced in this manner being known as coarse stuff and again should be stored for not less than 16 hours before use. It may be stored for a longer period but must be protected from drying out.

Both the above methods ensure that the maximum workability of the hydrated lime is fully developed.

Hydrated lime has almost completely replaced quicklime for site preparation of lime putty because of the time-consuming and labour-intensive operation of slaking quicklime. Some ready-mixed lime:sand for mortar manufacturers, however, still use quicklime as a starting material for lime putty for coarse stuff production.

If lime putty is made with semi-hydraulic lime there will be a slight loss of its hydraulic properties, but this will not be appreciable unless the putty or coarse stuff is stored for more than 3 days.

Dry mixing
Hydrated lime may be mixed dry with cement and sand, the necessary water added and used immediately. However, although the method has the advantage of greater convenience it does not give a mix with the same degree of workability as one incorporating lime putty.

Preparation of mortar mix
Gauging of coarse stuff and cement is sometimes not fully understood and can lead to disastrous consequences. To prepare a 1:2:9 mix, coarse stuff should first be prepared by mixing the lime putty and sand in the proportions of 1 volume of lime to $4\frac{1}{2}$ volumes of sand. The lime fills the voids or air spaces in the sand so that 1 volume of lime and $4\frac{1}{2}$ volumes of sand give $4\frac{1}{2}$ volumes of coarse stuff (not $5\frac{1}{2}$ volumes). Immediately before use, cement is added and if 1 volume of cement is mixed with 9 volumes of coarse stuff the resulting mix is 1:2:9.

If the only coarse stuff available is a 1:3 lime:sand, a 1:2:9 mix may be prepared by mixing 2 volumes of 1:3 coarse stuff with 1 volume of sand. This gives a 1:4$\frac{1}{2}$ lime:-sand coarse stuff. Immediately before use, cement is added in the proportion of 1 volume of cement to 9 volumes of the 1:4$\frac{1}{2}$ coarse stuff.

To prepare a 1:1:6 mix, coarse stuff is prepared from lime and sand in the proportions of 1 volume of lime to 6 volumes of sand. Immediately before use, cement is added in the proportion of 1 volume of cement to 6 volumes of coarse stuff. If a 1:3 lime:sand coarse stuff is used to produce a 1:1:6 mix, 1 volume should be mixed with 1 volume of sand. This gives a 1:6 coarse stuff. Immediately before use cement should be added in the proportions of 1 volume of cement to 6 volumes of 1:6 coarse stuff.

When the mortar is being made from coarse stuff, BS 5268: Part 3: 1985 recommends that about three-quarters of the required mixing water should be added to the mixer, followed by the required quantity of cement, which should be added slowly to ensure a thin paste free from lumps. The required quantity of coarse stuff should then be added and allowed to mix in, together with any additional water to achieve workability.

Once cement has been added to the mix the mortar should be used within 2 hours of the addition of the cement and under no circumstances should material which has commenced to stiffen be retempered or used.

2.3   Sands

The effect of sand on mortar strength and its working properties is considerable. Sands with excessive amounts of impurities, badly graded sands and poorly chosen sands can be responsible for serious defects in masonry. Time spent in specifying a suitable sand for a specific end use is never wasted – sand suitable for bedding masonry units may not be suitable for rendering and vice versa.

It is well known that a large proportion of building sand used in the UK, particularly that used for external rendering, does not conform to the appropriate limits prescribed in BS 1199 and 1200: 1976. However, despite their non-conformance with British Standard requirements many of the sands do perform in a satisfactory manner and undoubtedly further research is necessary in this area of construction.

Origin of sand

Natural sand is formed as the result of the gradual disintegration of rock masses due to weathering and the subsequent rolling and grinding together of the particles in rivers, on the sea-coast or by wind action. The most resistant of the common minerals composing rock masses is quartz, so that the larger remaining particles, i.e. the sand grains, will usually be composed mainly of this mineral. In most sands there will be a proportion of very finely divided material known as 'fines'. These usually consist of two different types of particles, the difference being due to the decomposing action of water on certain of the constituents of the original rock. Those unaffected by water will behave like minute sand grains and form the fraction of the 'fines' known as 'silt' and consist largely of quartz. Minerals such as felspars not only break up more readily than quartz, but in time are partially decomposed by water to form clay. Clay settles slowly in water and may travel considerable distances in suspension. Sometimes the clay and silt will be uniformly distributed throughout a deposit of sand, alternatively the 'fines' can vary considerably in different layers. Sometimes the clay forms a thin film on the surface of the sand grains, or occurs as nodules or balls irregularly distributed.

When sand has finally settled in a river, lake or sea, it is often covered by other materials from which salts, etc. may subsequently be deposited on the grains or in the pores. Many sands are inaccurately described as 'loamy', meaning the sand is 'clayey' or 'dirty'. Sands may be washed at the pit to remove these impurities.

Types of sand

Building sands can be divided roughly, as to source, into pit sands, sea sands, dune sands, dredged sands and crushed stone sands. In addition, crushed brick, clinker, or slag is sometimes used in lime and cement mortars.

Sands can also be distinguished broadly according to their grading (i.e. average particle size and distribution of particle sizes) and degree of cleanliness.

BS 1199 and 1200: 1976 define sand as follows:

Sand. A material mainly passing a 5.00-mm BS test sieve, which may be either a natural sand or one obtained by crushing hard rocks or gravels.

Natural sand. A sand produced by the natural disintegration of rock.

Crushed stone sand and crushed gravel sand. Sands produced, by crushing a hard stone or rock and a gravel respectively.

BS 1199 and 1200, since the issue of Amendment No. 1 in 1984, restricts the method of determining the grading of sands to that of washing and decantation followed by dry sieving as this method is believed to be more reproducible and to result in a more realistic measure of the particle size distribution of the sand. It also determines the clay and silt content which is now incorporated in the grading requirements.

Terms commonly applied to sands include:

Coarse. This contains more than about 20 per cent by weight of material coarser than 1.18 mm BS sieve, together with a relatively small proportion of material finer than 300 μm.

Fine. This contains more than 50 per cent of material finer than 300 μm.

Uniform. This contains mainly material of one size.

Gap-graded. These have some intermediate size missing, or present only in small quantity.

Smoothly, Regularly, or Well-graded. These have a well-distributed range of grain sizes.

Clean. This has a small amount of fine grains (i.e. particles readily washed through a 75 μm sieve).

Dirty. This contains too much silt and clay or organic contamination.

Clayey. This has a lot of 'fines', mostly clay.

Loamy. This has a lot of organic matter, i.e. humus from soil.

Soft. This is rather fine and clayey sand that may give particularly workable mixes.

Sharp. This has coarse, harsh, angular grains.

Effect of constituents on mortar

The various constituents which make up a sand have a marked effect upon the properties and subsequent performance of mortars and renders/plasters and initial consideration of these factors before specifying a sand frequently pays high dividends.

The coarse fraction. These particles usually have little effect on the setting of mortars, but crushed limestone particles or shell fragments may retard the set and reduce strength of certain anhydrous calcium silicate sulphate plasters which contain accelerators.

Silt. An appropriate proportion of silt improves the general grading and hence the workability of mortar mixes. Washed sand may be harsh-working, because of the absence of silt and an addition of sand containing suitable fines will considerably improve its working properties. Usually a high silt content in a natural sand is accompanied by an unduly high proportion of clay. Clean silt has no special effect on the setting of plasters, etc. although like all fine materials it accelerates the set of hemihydrate plaster.

Clay. This tends to retard the set of Portland cement mortars and is more critical in cement:lime:sand mortars because the cement is diluted by the lime and the setting time is consequently longer than for straight cement:sand mixes. Clay nodules may give rise to 'blowing' in renders exposed to the weather and frost, also pitting of exposed mortars. Excessive clay also increases the water requirements and consequently the drying shrinkage and ultimate strength of the plaster or mortar. A film of clay on the sand particles reduces the strength and adhesion of mortar but is an effective plasticizing agent, however more than 1 per cent lowers the early wet strength. BS 1199 and 1200 (1976) requires sands to be 'free from adherent coatings, such as clay, and from any appreciable amount of clay in pellet form'.

Inorganic impurities insoluble in water. Most sands, with the exception of white silver sands, contain iron compounds, usually in the form of iron oxide. Such compounds are usually harmless unless they form a persistent fine suspension when mixed with water which may affect the set of calcium sulphate plasters.

Iron pyrities. When present, this may give trouble through oxidation and expansion, or causing staining. Large proportions of mica, shale and other laminated materials are examples of harmful impurities.

Calcium carbonate. This is frequently present but generally only in small quantities and has no effect other than those previously mentioned.

Inorganic impurities soluble in water. The quantity of soluble salts in sand is rarely sufficient to seriously affect the set of mortar or plaster but may accelerate the corrosion of unprotected metalwork. Sea or esturine sands are frequently responsible for dampness, efflorescence and corrosion when used for mortar or plaster. SCAG (British Ceramic Research Limited, 1988) limits the chloride iron content in sand to 0.15 per cent by mass of the cement in any specified mortar mix.

Organic impurities

Humus. The setting of Portland cement mortars and renders may be retarded or even prevented by the presence of this impurity in sand.

Coal. Soft lignites and some bituminous coal particles may be responsible for popping, pitting or blowing in sanded plasters.

Organic slimes and industrial wastes. Sands containing these contaminates may affect the set of plasters and mortars.

Grading of sand

The ideal grading of sands for mortar is one approaching a uniform proportion (by weight) of particles in each of the various ranges of sizes which result from the use of the specified sieves and including in addition some silt and a small proportion of clay as a plasticizing agent.

When mortar is made from a well-graded sand the smaller grains will tend to fit in between the larger ones, leaving smaller voids to be filled with cement and lime resulting in a good workable mix. If mortar is made with a uniform coarse sand the voids between the grains are relatively large and it is quite likely that there will not be sufficient cement and lime in the mix to fill them.

Uniform sands may also cause problems and their working properties can be improved by mixing with clean coarse sand.

For rendering a well-graded sand is particularly important as badly graded sand is difficult to work. BS 1199 and 1200 (1976) specifies the grading requirement for sands for mortar for plain and reinforced brickwork, blockwalling and masonry, also for external renderings, internal cement and lime plastering.

Figure 2.1 shows the specified grading limits for Type 'S' sands and Type 'G' sands (broken lines) based on Table 1 of BS 1200: 1976: Amendment No. 4510. The Figure also shows Type 'M' sands (dotted lines) to BS 882: 1983.

Either Types 'S' or 'G' can be used for masonry mortars. However, BS 5628: Part 3: 1985 warns that 'Not all sands complying with BS 1200 will be suitable for conditions of Severe or Very Severe exposure or where flexural strength (adhesion) is critical, owing to high fines content and/or particle distribution. In such cases, consideration should be given to using sands having a particle size distribution towards the coarser end of the BS 1200 (1976) grading envelope. Such sands may be found among those complying with grade 'M' or BS 882: 1983.

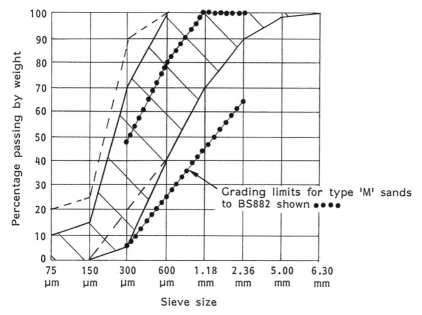

Figure 2.1 Specified* grading limits for Type 'S' sands. (Type 'G' sand limits indicated by – – – – ; type 'M' sands indicated by ••••). * BS 1200: 1976 AMD. 4S10, 31 May 1984

Simple site tests

On small building sites where laboratory facilities may not be available, simple site inspection and testing of sands are frequently worthwhile and can be extremely cost effective in that such checks will confirm the suitability (or otherwise) of any particular unfamiliar sand.

Check on harmful particles. Examination of a small amount of sand with a magnifying glass should reveal the presence of undesirable impurities such as coal dust, balls of clay, etc.

Dirty sand. Initial testing of the sand can be carried out simply by rubbing some sand between the fingers. If it balls easily or leaves a stain further laboratory tests are advisable.

Silt test. Using a clean jam jar or similar vessel with a screw top lid, three-quarters fill it with a saline solution (one teaspoon of salt to a half-litre of clean water). Half fill the jar with a representative sample of sand, replace the lid and thoroughly shake the contents. After 3 or 4 hours a layer of silt should have formed on top of the sand. The total depth of silt and clay should be compared with the depth of the sand, if it exceeds 10 per cent the sand is too dirty to use. If the water retains a red cloudiness this may be due to an excessive quantity of iron oxide.

Bulking of sand

When sand becomes damp each grain is surrounded by a film of moisture which prevents complete physical contact, so producing an increase in bulk volume known as bulking. The amount of bulking for a given sand varies with the moisture content, being maximum when every grain is just moist over its complete surface. If not all the grains are coated, the increase in volume is less whereas if more moisture is present than is required just to coat each grain, the surface tension between the particles is reduced and bulking decreases. When sand is saturated its volume is equal to its dry volume, or even slightly less due to the complete breakdown in surface tension.

When mortar mixes are proportioned by volume, the effect of bulking is to produce cement rich mixes. Fine sands bulk more readily than coarse sands because of their greater surface area and the better the grading of any sand the greater is the volume increase when moist. If volume batching is to be used it is good practice to test the sand for bulking and if necessary adjust the mix proportions of the mortar accordingly.

Bulking test. A simple test for bulking of sand can be carried out as follows:

Use a straight-sided glass jar (preferably a graduated measuring cylinder – Figure. 2.2) and pour in the sample of sand to be tested, without compaction until the jar is approximately two-thirds full and level the specimen. Measure the depth of sand, either from the graduations on the jar or by pushing a rule through the sand to the bottom of the jar. Empty the sand into another clean container taking care not to lose any and half fill the original jar with clean water. Replace half the sand and rod it thoroughly to remove air, then add the remainder, rod again and level off and measure the new depth. The percentage bulking can then be calculated, i.e.:

$$\frac{Original\ depth\ of\ sand\ -\ wet\ depth\ of\ sand}{wet\ depth\ of\ sand}$$

The volume of sand used should then be the amount specified plus the percentage bulking. The bulking test should be carried out at the start of the work, after each delivery of sand, and after any change in the weather likely to affect the moisture content of the sand appreciably. When batching by weight dampness barely affects

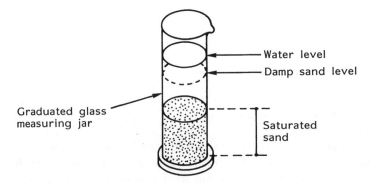

Figure 2.2   Simple test for bulking of sand

the proportioning and no adjustments are necessary hence the trend towards weight batching.

2.4   Admixtures

An admixture may be defined as a material, other than water, aggregate, lime and cement which is added to a mortar or grout during mixing to modify some property or properties of the mortar or grout. The term 'admixture' is used in preference to the alternative 'additive' which is reserved for materials used by cement manufacturers to modify the properties of cement. Some materials do, of course, justify both descriptions. Admixtures for concrete have been classified under the following 15 groups:

1 Accelerating; 2 Water-reducing, and set controlling; 3 Grouting; 4 Air-entraining; 5 Air-detraining; 6 Gas-forming; 7 Expansion producing; 8 Finely-divided mineral; 9 Damp-proofing and permeability-reducing; 10 Bonding; 11 Chemical admixtures to reduce alkali-aggregate expansion; 12 Corrosion-inhibiting; 13 Fungicidal, germicidal and insecticidal; 14 Flocculating; and 15 Colouring.

It should be appreciated that several admixtures contain materials which separately would belong in two or more of the foregoing groups.

Air-entraining admixtures

Air-entrainment deliberately introduced into a mortar in the form of a relatively stable system of small air bubbles, increases the workability of the mortar. High air content may reduce the bond between unit and mortar but gives higher resistance to frost in early life. The plasticizing agent is usually added to the mixing water as a concentrated liquid which, if used too liberally or overmixed can result in a mortar overaerated with danger of grossly inadequate strength (BDRI, 1967).

It is possible, in fact, to produce workable mixes with unacceptably high sand:cement ratios with consequent danger of very much lower mortar strengths. Some authorities state that the reduction of tensile bond between unit and mortar which occurs with increasing air content can be offset by the use of admixtures of methyl cellulose, but not all research bears this out. When air-entraining agents are used, it is important that the mixing of the mortar should be carried out in a cement mixer or by hand, as the use of a power mill may eliminate a high proportion of the entrained air, resulting in a harsh mix. Early work by Powers (Powers, 1945, 1949; Powers and Helmutt, 1953) in the USA suggests that entrained air may prevent damage by freezing in hardened cement paste because destruction of the paste is caused mainly by hydraulic pressure generated due to the freezing of the water rather than by direct crystal pressure produced through the growth of bodies of ice crystals. Development of this hydraulic pressure is a consequence of the expansion of approximately 9 per cent which accompanies the conversion of water to ice, because this expansion tends to press the unfrozen water from the freezing zone through capillary pores. As a result, tensile stresses are set up in the hardened paste which cause failure if they exceed the tensile strength of the paste. The magnitude of the hydraulic pressures and of the tensile stresses are said to be dependent on several factors. The most important is, perhaps, the length of capillary pores, because the pressure is proportional to this length. The length of capillary voids can be reduced by a satisfactory distribution of voids of entrained air, resulting in a relief of hydraulic pressure during progressive freezing. This means that the air voids are more effective the closer they are together. That is, for a given air content, the protection afforded by the voids against damage by freezing and thawing is usually greater when the size of the voids is decreased, or when the number of voids per unit volume of paste is larger.

Plasticizers should conform to the requirements of BS 4887: 1973 and only be used when specified by the designer.

Domestic detergents should never be used as a substitute for mortar plasticizers as although they may be perfectly satisfactory from a workability point of view they are sometimes responsible for unsightly staining of walls when some of their constituents are leached out by rainwater.

Expansive admixtures

Finely divided iron with ammonium chloride is used to produce some expansion in grouts and mortars by oxidation. These materials are used for machine grouting and similar purposes.

High-alumina cement and calcium sulphate or calcium sulphoaluminate added to Portland cement in predetermined amounts produce self-stressing or expansive cements. Such cements have been developed in the USA, Japan and the USSR but are not available in the UK.

Commercial applications appear to be limited by the high temperature sensitivity of the chemical reactions which take place and by other difficulties.

Bond and abrasion resistance admixtures

Several natural and synthetic polymer emulsions and dispersions are available for incorporation in mortar or concrete to improve the abrasion resistance of wearing surfaces. They may also improve the bond strength, decrease the modulus of elasticity and increase the tensile strength. Some decrease in compressive strength is generally noted and an increase in the drying shrinkage. Natural rubber latex, polyvinyl acetate emulsions, styrene–butadiene latex, acrylic, divinylidene and polyvinyl chloride emulsions are used for this purpose.

Formulations may include emulsifying agents, anticoagulants, anti-oxidants, flexibilizers, air detraining and other additions according to the nature of the materials and the intended purpose.

Natural rubber latex. This is noted for the high bond strength which develops when incorporated in mortar but it age-hardens and is sensitive to oil and grease.

Styrene–butadiene artificial rubber latex. This is noted for good chemical resistance but develops rather lower bond strength than natural rubber.

Polyvinyl acetate. This is highly compatible with Portland cement, but the film is sensitive to moisture and the bond strength is impaired in permanently damp exposurers. Acrylic emulsions are alternative materials which are more expansive but less sensitive to moisture after drying; they probably develop less bond strength. Polyvinyl chloride latex is not generally available in the UK as an admixture.

Mineral powder plasticizers

Finely ground powders are sometimes added to mortar with the object of improving the workability and plasticity of the mix. They can be used to supplement the cement in mixtures deficient in fine material. Some powders such as ground limestone, whiting, ground sand, hydrated lime or bentonite are relatively inert chemically. The finely divided minerals referred to will generally have an average fineness similar to or greater than that of ordinary Portland cement.

Mineral plasticizers in mortar may be responsible for a reduction in water demand compared with mortar of equal workability without the admixture, due to the physical characteristics of the powder. On the other hand, the water demand may be increased in which case the shrinkage and permeability of the hardened mortar may be increased.

Fungicidal admixtures

Copper sulphate (0.1 per cent), penta-chlorophenol (0.2 per cent) and other compounds have been suggested as admixtures to deter the growth of algae, lichens, etc. on hardened mortar. In general the fungicides have a limited life and tend to wash out of the mortar, losing effectiveness with time. With some, the problem of toxicity to human or animal life is an important factor.

Accelerators and frost inhibitors

Accelerators and frost inhibitors based on calcium chloride should never be used in mortar. However, some manufacturers of proprietary frost inhibitors do include calcium chloride in their products. It is also used by some manufacturers of extra rapid hardening Portland cement and masonry cement. The effect of additions of calcium chloride to mortar mixes is to increase the rate of development of strength and resistance to erosion and abrasion but this is unfortunately accompanied by high shrinkage and possibly creep, as well as a greater tendency to cracking. Calcium chloride is deliquescent and the possibility of corrosion of steel wall ties and reinforcement embedded in mortar is a subject of controversy. On the question of frost inhibitors BS 5628: Part 3: 1985 states 'Although when frost conditions are anticipated there would be some advantage in accelerating the setting of the mortar, in practice no suitable admixtures are known that are free from other undesirable effects. In particular, calcium chloride or admixtures based on this salt may lead to subsequent dampness or corrosion of embedded metals, including wall ties, and therefore should never be used. There is little experience of the successful use of any admixture intended to provide frost protection by depressing the freezing point of the mixing water. Some substances that might be contemplated for this purpose, e.g. ethylene glycol, are known to adversely affect the hydration of the cement'.

Retarders

Some ready-mixed cement:lime:sand mortars contain a retarder that enables them to be used for extended predetermined periods after delivery. Retarders generally slow down the hardening of the mortar and should not be confused with the addition of salts which may speed up the setting but inhibit the development of strength. Materials known to exhibit a retarding action are sugar, carbohydrate derivations, soluble zinc salts and soluble borates, etc. Some retarders commonly used in practice are also water-reducing, the two main groups being (i) lignosulphonic acids and their salts and (ii) hydroxylated carboxylic acids and their salts. It is also necessary to exercise great care when using retarders as incorrect quantities or uneven distribution in the mortar can seriously affect the setting and hardening processes. It is essential that trial mixes be made, particularly when lignosulphonic acids are used as a greater degree of air entrainment is likely, although this may vary considerably depending upon the constituents of the mortar.

Retarding agents if accepted, should be very carefully selected as even though the set is retarded, the rate of loss of workability does not necessarily follow the same pattern.

Pigments

It is essential that pigments used for colouring mortar should have: fastness to lime and cement; fastness to light; resistance to aggressive chemicals, such as waste gases in the air and harmful constituents of special building materials; temperature stability; high tinting strength and freedom from materials such as acids or soluble salts which may react with the cement and cause efflorescence. In addition pigments must not affect the durability, mechanical strength of the cement or tensile bond of mortar to unit; their particle size, wettability, etc. must be such that they mix readily and evenly with the cement.

When the pigment is to be applied by mixing it into the mortar, the full value of the colour is only obtained by thorough mixing. BS 1014: 1975 specifies pigments for use in Portland cement, giving the composition and requirements of each pigment, and methods of test including analytical procedures for the main constituents. Pigments which may be satisfactory as a surface coating may be quite unsatisfactory when incorporated in cement or lime mixes.

Although inorganic pigments are more permanent than organic pigments, having greater resistance to lime, cement and chemical attack and are unaffected by light, they can still cause problems. It has been shown that the compressive strength of mortar is affected when increased quantities of pigment were required to give strong colours. As early as 1921, Abrams determined the compressive strength of 1:2 cement:sand mortar containing a blue colouring material, red oxide up to 15 per cent (in terms of weight of cement) and carbon black up to 10 per cent, and found the blue caused a pronounced increase in strength, the red a slight decrease and the carbon black a material decrease. The writer (Thomas, Coutie and Pateman, 1970) observed a decrease in compressive strength of over 30 per cent when a pigment containing carbon black was added in proportions of up to 10 per cent in a $1:\frac{1}{4}:3$ cement:lime:sand mortar. The decline in strength was found to be approximately linear with 2 per cent increments of pigment up to 10 per cent for each of the 7,14 and 28-day series of specimens tested.

BCRL (1988) recommend that carbon black should be limited to 3 per cent by weight of the cement for structural brickwork. The colour retention of cement products containing carbon black may be variable when exposed externally and the Standard (BS 1014: 1975) recommends that users of pigments for outside use obtain suitable assurances, additional to the safeguards provided by the specification, before carrying out any large amount of work with types of carbonaceous black with which they are not familiar.

2.5   Water

The quality of water used for gauging mortar is important since contaminated water may lead to impaired performance of the hardened mortar. As a general rule, if the water is suitable for drinking it is clean and free from harmful impurities and suitable for making mortar. Where the quality of the supply is doubtful it should be tested in accordance with BS 3148: 1980.

Sea water should never be used for making mortar.

2.6   Mortars for brickwork and blockwork

General

The desirable properties of all mortar mixes for masonry (i.e. brickwork, blockwork and natural stonework) are said to be (a) workability; (b) good water retentivity; (c) sufficiently early stiffening; (d) development of suitable early and final strength; (e) good adhesion or bond; (f) durability; and (g) where appropriate the necessary aesthetic properties. Modern mortars usually consist basically of cement, lime, sand and mixing water although lime is often replaced or used in conjunction with a plasticizer. Lime and cement mainly determine the rheological properties of mixes (the workability) although these may be modified by the grading of the sand or the effect of additives. Workability, water retentivity and bond are properties which lime imparts, while the cement and sand confer strength and durability. The use of all admixtures should be strictly in accordance with the manufacturer's instructions as too little, or perhaps more important, too much of the additive may produce disastrous results. For example, when pigments are used they tend, due to their non-cementitious character and relatively high surface area, to adulterate the mix, reducing bond strength and in some instances compressive strength (particularly carbon black pigment).

Plasticizers of the air-entraining type improve frost resistance during laying and aid workability but if over prescribed, particularly with high-suction units, may be the cause of poor bond. When masonry cements are used, it is important to realize that in the UK proprietary brands contain about 75 per cent Portland cement and 25 per cent of a fine filler such as ground limestone and an air-entraining agent. Allowance should therefore be made for the reduction in cement content when specifying mortar mixes based on masonry cement. In other countries, the percentage may be higher or even considerably less.

Calcium chloride and frost inhibitors based on calcium chloride should not be used in mortars, as apart from being ineffective such additives cause deliquescence and an increased risk of corrosion to ferrous metals when present.

TABLE 2.1 Ready-mixed lime:sand mixes for specified cement:lime:sand mortars. (Based on BS 5628: Part 3: 1985.)

Mortar designations	Type of mortar – specified cement: lime:sand mortar (proportions by volume)	Lime:sand mix (proportions by volume)	Gauging of cement with lime:sand mix (coarse stuff)
(i)	1:0 to $\frac{1}{4}$:3	1:12	1:3
(ii)	1:$\frac{1}{2}$:4 to 4$\frac{1}{2}$	1:9	1:4$\frac{1}{2}$
(iii)	1:1:5 to 6	1:6	1:6
(iv)	1:2:8 to 9	1:4$\frac{1}{2}$	1:9
(v)	1:3:10 to 12	1:4	1:12

Ready-mixed lime:sand for mortar (Table 2.1) is an excellent basis for mortar as it is subject to factory control and only needs to be gauged with cement before use. Ready-to-use retarded cement:lime:sand mortars and cement:sand mortars may be suitable for non-structural use where properties such as consistent colour are particularly required but they are not generally suitable for structural masonry. Readers are referred to BS 5628: Part 1: 1978, Part 2: 1985, Part 3: 1985, PD 6472: 1974 and BS 4721: 1981 for further information.

Mortar mixes

Designers should select a suitable mortar designation and type after taking into account the type of construction, position in the building, degree of exposure and the possibility of early exposure to frost, together with the general properties of mortar given in Table 2.2.

TABLE 2.2 Mortar mixes. The direction of change in properties is shown by arrows

		Mortar designation	Mortar type		
			Cement:lime: sand (proportions by volume)	Air-entrained mixes	
				Masonry cement: sand (proportions by volume)	Cement: sand with plasticizer (proportions by volume)
↑ Increasing strength and improving durability	Increasing ability to accommodate movements due to temperature and moisture changes	(i)	1:0–$\frac{1}{4}$		
		(ii)	1:$\frac{1}{2}$:4–4$\frac{1}{2}$	1:2$\frac{1}{2}$–3$\frac{1}{2}$	1:3–4
		(iii)	1:1:5–6	1:4–5	1:5–6
		(iv)	1:2:8–9	1:5$\frac{1}{2}$–6$\frac{1}{2}$	1:7–8
	▼	(v)	1:3:10–12	1:6$\frac{1}{2}$–7	1:8

Increasing resistance to frost attack during construction. →
Improvement in adhesion and consequent resistance to rain penetration. ←

Ready-mixed lime:sand for mortar

Table 2.3 gives the mix proportions for ready-mixed lime:sand for mortar which may be used if the manufacturer's recommendations are not available.

The range of volume proportions given in Tables 2.2 and 2.3 is to allow for the effect of the differences in sand grading upon the properties of the mortar. Weight batching produces more consistent mortars than volume proportioning, provided that the variation in bulk densities of the materials is checked regularly.

TABLE 2.3 Ready-mixed lime:sand for mortar

Mortar designation	Specified cement: lime:sand mortar (by volume)	Category of lime:sand mix (by volume)	Amount of cement to be gauged with lime:sand mix (by volume)	
			Normal use	Special use
(i)	1:$\frac{1}{4}$:3	1:12	1:3	–
(ii)	1:$\frac{1}{2}$:4–4$\frac{1}{2}$	1:9	1:4$\frac{1}{2}$	1:4
(iii)	1:1:5–6	1:6	1:6	1:5
(iv)	1:2:8–9	1:4$\frac{1}{2}$	1:9	1:8
(v)	1:3:10–12	1:4	1:12	1:10

Table 2.4 gives the mix proportions for mortar based on weight.

TABLE 2.4 Composition of mortars by weight

Mortar designation	Cement:lime:sand mortar	Masonry cement:sand mortar	Cement + plasticizer:sand mortar
(i)	1:0.1:4	–	–
(ii)	1:0.3:6	1:3.5–4.5	1:4–5
(iii)	1:0.45:8	1:5–6.5	1:6.5–8
(iv)	1:0.9:12	1:7–9	1:9.5–10.5

Mortar for structural masonry

When mortar is used for structural masonry its strength properties become even more important and if 'special category' masonry is used higher permissible stresses are allowed and compressive strength testing of the mortar becomes necessary (see BS 5628: Part 1: 1978). The strength requirements of the mortar cubes or prisms are shown in Table 2.5.

TABLE 2.5 Mortar strength requirements

Mortar desugnation	Mean compressive strength at 28 days (N/mm^2)	
	Preliminary (laboratory) tests	Site tests
(i)	16.0	11.0
(ii)	6.5	4.5
(iii)	3.6	2.5
(iv)	1.5	1.0

Selection of mortar grades and their performance depends largely on the location of the masonry and guidance on the selection of units and suitable mortars are given in Chapter 3.

2.7   Mortars for natural stone masonry

General

The selection of the correct mortar is extremely important as not only does it affect the strength and appearance of the wall but also its long-term durability and weather resistance (Table 2.6). The mortar should be durable but not stronger than the stone; it should be easily workable; have good water retentivity; sufficiently early stiffening; development of suitable early and final strength; good adhesion or bond; appropriate aesthetic properties and the ability to accommodate minor settlement in the wall.

Factors that need to be considered are as follows:

Type of stone

Mortar for stone usually consists of hydraulic lime or more usually cement:lime mortar. The former is sometimes used for very porous stone in the proportions 2.5 hydraulic lime:sand and the latter in designation (v) mortar, i.e. 1:3:12 Portland cement:lime:sand. If greater strength is required (i.e. for copings or projections) a designation (iv) mortar, i.e. 1:2:9 Portland cement:lime:sand would be more appropriate. These mixes should also be used for flint walling. For some sandstone and granite a stronger mortar may be used such as designation (iii), i.e. 1:1:6 Portland cement:lime:sand or for very dense granite a designation (i) mortar, i.e. $1:0\frac{1}{4}:3$ Portland cement:lime:sand.

The weaker cement mortar described above may not be satisfactory for use in cold weather and it is recommended that if possible the work be postponed to avoid frost damage. If this is not possible it may be necessary to use a stronger mortar or to include a plasticizer.

Type of construction

In ashlar walling and rubble walling the mix proportions are likely to be similar but the type of fine aggregate may vary. For ashlar the mortar is usually finer and may include crushed stone rather than sand to give a more compatible effect.

Rubble walling, being more irregular usually has thicker joints and the aggregate should be coarser and may consist of grit as well as sand or crushed stone. The choice is largely a matter of aesthetics and local tradition. However, crushed oolitic limestone is unsuitable as crushing of the ooliths destroys their strength and produces a smooth texture.

Type of aggregate

The aesthetics and strength of the wall are affected by the choice of sand/fine aggregate used in the mortar mix. Soft sands with worn and rounded grains make the mortar easily workable but are of inferior strength to mortar gauged with sharp sand which has angular grains and consequently poorer workability. Sharp sand

TABLE 2.6 Summary of mortar types and designations for natural stone masonry

Mortar designation	Mortar type (proportions by volume)					
	Cement:lime: sand	Masonry cement:sand	Cement: sand with plasticizer	Hydraulic lime:sand	Lime:pulverized fuel ash:sand	Lime:brick-dust:sand
(i)	1:0 to $\frac{1}{4}$:3	—	—	—	—	—
(ii)	1:$\frac{1}{4}$ to $4\frac{1}{2}$	1:$2\frac{1}{2}$ to $3\frac{1}{2}$	1:3 to 4	—	—	—
(iii)	1:1:5 to 6	1:4 to 5	1:5 to 6	—	—	—
(iv)	1:2:8 to 9	1:$5\frac{1}{2}$ to $6\frac{1}{2}$	1:7 to 8	—	—	—
(v)	1:3:10 to 11	1:$6\frac{1}{2}$ to 7	1:8	2:5	—	—
(vi)	—	—	—	2:5	—	—
(vii)	—	—	—	1:3	2:1:5	—
(viii)	—	—	—	—	—	2:2:5
(ix)	0:1:1	—	—	—	3:1:9	—
(x)	—	—	—	—	—	1:1:3
(xi)	0:2:5	—	—	—	—	—

Increasing strength and durability ↑

Increasing ability to accommodate movement, e.g. due to settlement temperature and moisture changes ↓

may occur naturally or be produced from crushed stone. While sand can be obtained in various colours depending on the locality and geological nature of the source, pigments can be added if necessary. When pigments are used they tend, due to their non-cementitious character and relatively high surface area, to adulterate the mix, reducing bond strength and in some instances, compressive strength (particularly carbon black pigment).

Black ash mortar is not recommended generally and specifically if steel wall ties, etc. are to be used.

Jointing and pointing

Ashlar stone masonry should be bedded and pointed with the same mortar as the work proceeds to ensure a homogeneous joint. If coloured pointing is used the specially prepared mortar should be of an identical mix to the bedding mortar. The bed joint should be raked out as the work proceeds to a depth of 20 mm and later pointed with the coloured mortar. Pointing of ashlar work is normally finished flush with the face of the wall.

Rubble masonry may be jointed and pointed simultaneously or pointed as for ashlar masonry in a special mortar. The character of rubble walling is more suited to simultaneous jointing and pointing and a variety of techniques are available. A coarse texture to the very slightly recessed joint can be achieved by stippling the surface of the mortar with a stiff bristle brush before it finally sets in order to expose the grit in the mix.

Ribbon point projects from the face of the wall and is finished with a trowel. This is generally carried out as a separate operation after the wall is built. If a dense mortar is used this type of pointing may accelerate the decay of the stone. Also, it can cause water to lodge against the stone until eventually frost action causes the mortar to dislodge. This is a method of pointing to be avoided for external work.

If rubble masonry is very irregular the joints are sometimes filled with small stones or flints (galleting) of different shapes depending upon the type of stone (BS 5390: 1976).

2.8   Mortar joints

Joints

The average thickness of both the horizontal and vertical mortar joints is dictated by the co-ordinating size of the masonry units and is normally taken to be 10 mm. This joint size allows for irregularities in the masonry units and should accommodate most oversize particles in the sand while being reasonably economical in the use of mortar. Alternative joint thicknesses may be specified for numerous reasons but care needs to be taken not to impair the structural integrity of the resulting construction.

Pigments are often used in mortars to provide a contrast to, or blend with, the brick colour. When used care should be taken to ensure that the manufacturer's instructions are strictly complied with. Over dosage can have deleterious results as previously discussed. Where carbon black is used as a colouring agent quantities greater than 3 per cent by weight of the cement may affect the bond and strength characteristics of the masonry. When other pigments are used the quantity should generally be limited to a maximum of 10 per cent by weight of the cement. To achieve maximum consistency of colour and strength a ready-mixed lime:sand mortar should be used as the pigment is added under controlled factory conditions.

Jointing and pointing

Mortar joints may be finished in a number of ways. When this is carried out while the mortar is still fresh it is termed jointing. When the mortar is allowed to stiffen and some then removed and replaced with fresh mortar, sometimes coloured, before finishing, the process is referred to as pointing. Pointing reduces the area of mortar carrying the load from one course to the next and therefore must never be used in loadbearing masonry without first having checked the calculations for stress and slenderness ratio. Where pointing is carried out it is essential for the mortar used in the pointing to be the same strength or weaker than that used in the joints. If the pointing mortar is stronger than the bed mortar spalling of the masonry units may occur due to the load being transferred to the face of the unit via the stronger mortar.

It is also important to avoid pointing over d.p.c.s as this can provide a passage for water to bypass the d.p.c. and can cause mortar crumbling as the d.p.c. settles. In extreme cases the masonry faces may be sheared off.

Joint profiles

The appearance of the masonry will depend a good deal upon the shape of the joint. Five types are illustrated. There is some argument as to the respective merits of finishing the joint as the work proceeds as against raking and pointing. Probably finishing as the work proceeds will give a joint rather more resistance to frost but it may be difficult to keep the colour consistent throughout a large area of wall. If raking and pointing is carried out then the raking must be very thorough and should be to a depth of approximately 20 mm. The type of joint chosen needs to be related to the kind of unit used.

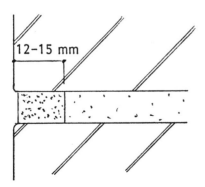

12–15 mm

Figure 2.3 Flush or bag-rubbed joint. This finish gives maximum bearing area and is often favoured when coarse textured units are used. With some brick types the finish may appear a little irregular.

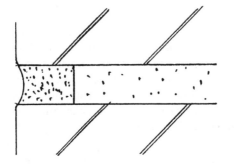

Figure 2.4 Curved recessed (bucket handle) joint. This joint can give an improved appearance over a flush joint with negligible reduction in strength. It is generally considered that this joint gives the best weather resistance due to the smoothing of the joint and the superior bond this achieves. It is perhaps the most commonly used joint.

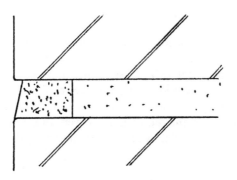

Figure 2.5 Struck or weathered joint. Weathered bed joints produce an interplay of light and shadow on the masonry. Such joints when correctly made have excellent strength and weather resistance.

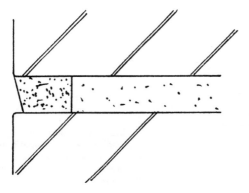

Figure 2.6 Overhung struck joint. This gives a slightly different appearance of light and shade to struck or weathered jointing. Unfortunately, it allows rainwater to lodge on the horizontal faces of the units and thus to penetrate the bricks and/or joints causing discolouration and possible frost damage. For these reasons it should be confined to lightly stressed interior walls and external masonry using units known to have high durability and if clay, low soluble sulphate content.

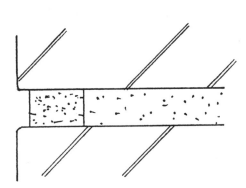

Figure 2.7 Square recessed joint. The square reccessed joint when used with durable units, can produce a very pleasing effect but its weather resistance and strength will be considerably less than struck flush or curved recessed joints. With heavily perforated units where the perforations occur near to the face a recessed joint may be inadvisable because resistance to water penetration may be impaired. **NOTE**: Some units are unsuitable for overhung struck and square recessed jointing because of their texture and relatively low frost resistance.

2.9   Re-pointing and repair of old brickwork

Before re-pointing old brickwork it is very important that some thought should be given to the treatment of the joints to ensure that they are compatible with the texture and colour of the bricks. If the wrong profile, colour or procedure is used this may deface a fine old brick wall. Not only can the unsympathetic treatment of a wall downgrade it, selection of the wrong materials for the job may even be responsible for damaging the old bricks. Assessment of the properties of the original bricks and mortar should always be made before deciding upon a suitable mortar mix for the remedial work. It goes without saying that the standard of workmanship can make or mar the end result.

Mortar mixes

The mortar for old brickwork generally consists of lime mortar, i.e. a mix of hydrated lime and sand which did not require gauging with cement. The mix proportions usually being 1:2–3 hydraulic lime:sand. It is unlikely that a lime mortar will be used for re-pointing work but to use a strong pointing mortar in conjunction with a lime bedding mortar is asking for trouble. The latter would tend to act as a knife edge and as differential movement occurs between the hard pointing and the softer bedding mortar there is a danger with low-strength bricks that spalling of the face will occur.

Mortar mixes should be carefully selected to suit the existing bricks and the exposure conditions. Under normal conditions of exposure a 1:1:6 Portland cement:lime:sand mix is generally acceptable. For the low-strength bricks such as Stocks and some handmades, a 1:2:9 mix is usually satisfactory and relatively compatible with lime mortars. The higher lime content mortars are less rigid and therefore more capable of accommodating thermal, moisture and other movements and consequently may provide superior resistance to rain penetration. For very hard dense bricks used in situations of extreme exposure, a $1:0 - \frac{1}{4}:3$ Portland cement:lime:sand should be used but with the richer cement mixes more shrinkage of the joints is likely to occur resulting in hair-line cracking at the brick/mortar interface.

Soft, washed sand is generally used for pointing and silver sand is frequently preferred.

If coloured mortars are used for pointing the pigments should comply with BS 1014: 1975 and should not be trowelled excessively as this can affect the colour achieved. Once a mix has been established producing an acceptable colour, nothing should be changed. Altering the proportions or the source of the supply of the sand or cement would almost certainly alter the colour of the mortar.

Old lime mortars tend to be lighter in colour than modern mortars based on Portland cement and pigments should only be used if satisfactory colour matching cannot be achieved by using natural materials, such as lime with the minimum amount of cement.

If a lime mortar is used (i.e. without an addition of cement) the lime should be in the form of lime putty. However, it should be remembered that lime is caustic and should always be added to water and *not* vice versa. Quicklime is especially dangerous and special precautions should be taken when handling this material such as the use of safety goggles, face masks and gloves, etc.

Methods of preparing lime putty on site using non-hydraulic and semi-hydraulic lime or quicklime are given on p. 32 and elsewhere (BS 6270: Part 1: 1985). Pozzolanic constituents such as pulverized fuel ash (PFA) or finely ground clay bricks can be used to enhance the hydraulic strength of lime:sand mortars but it is essential that the sulphate content of the PFA should not exceed 1 per cent mass.

Pozzolanic constituents should be added to the mix only just before use.

Workmanship

Re-pointing should only be entrusted to experienced masons, since removal of the existing mortar involves risk of damage to the arrises and particular care is required when dealing with ornamental masonry.

When re-pointing small areas of large walls, the finish should match that of the surrounding masonry, except that it may be left slightly recessed to minimize the contrast with original work. Prior to the work of re-pointing, all traces of lichen and moss should be removed by careful scraping and if necessary by chemical treatment, care being taken not to damage the face of the masonry.

Raking out of the original mortar should be to a depth of not less than 12 to 15 mm to provide an effective key for the new mortar. Care should be taken not to damage arrises, especially where the old mortar is hard and well bonded and particular care should be taken with fine joints.

Removal of mortar joints can be carried out with cutting discs (frequently large discs are used for the bed joints and smaller discs for the perpends) and careful use of such tools is much preferred to chisels which may cause damage due to impact. After removal of the necessary amount of the existing mortar joints, the joints should be cleaned out thoroughly and wetted immediately before filling. When filling the joints the mortar should be pressed well in to ensure complete penetration and bond to the units and mortar. Tooling of the joints to compact the mortar helps to improve the durabilities of the mortar and the rain shedding capacity of the wall. However, the mortar should never be allowed to spread over the face of the masonry. The surface of the pointing should be kept slightly back from worn arrises to prevent a thicker joint being expressed on the face of the wall and the creation of feather edges. Re-pointing should be carried out from the top of the wall downwards.

In some instances mortar samples are taken and sent for chemical analysis to check the mix proportions of the original jointing. When this is carried out the recommendations for sampling and testing in BS 4551: 1980 should be followed.

Mixing, proportioning and batching of mortars should be in accordance with the requirements of BS 5628: Part 3: 1985.

Sulphate action

Where sulphate action has occurred repair may involve complete rebuilding of the walls. However, if the trouble is confined to relatively minor damage of the joints every effort should be made to exclude the source of water in the construction before repairing the jointing. Re-pointing should then be carried out using a sulphate-resisting cement in the mortar.

When not to re-point

Re-pointing should not generally be carried out during the winter period when there is danger of frost action occurring. However, if it does become necessary to re-point in winter, the work should be protected against the effects of icy winds and rain.

Re-pointing should not be used as a means of shrouding decay of the mortar due to (a) the ingress of water due to faulty damp-proof course details at ground level; cills; lintels; parapets and roof junctions, etc. (b) frost damage due to the use of unsuitable mortars; (c) severe sulphate attack of the mortar joints or when appropriate of the backing materials to the masonry; (d) permanent damp situations due to the use of inappropriate units; (e) water damage from leaky gutters, rainwater downpipes or overflow pipes; and (f) structural movements. All such problems should be investigated and remedied before re-pointing.

References

Abrams, D. A. (1921) Effects of colouring materials on the strength of mortar. *Eng. News-Record*, **86**, 721.
BDRI (1967). *No substitute for lime*. Nos 35 & 36 Newsletter, Brick Development Research Inst., Melbourne. Jan and March.
British Ceramic Research Limited. The 1988 SP 56 *Model Specification for Clay and Calcium Silicate Structural Brickwork*. Building Materials Division.
BS 12: 1978 *Specification for ordinary and rapid hardening Portland cement*. BSI, London.
BS 146: Part 2: 1978 *Specification for Portland–blastfurnace cement*. BSI, London.
BS 5628: Part 3: 1985 *Use of masonry – Materials and components, design and workmanship*. BSI, London.
BS 4027: 1980 *Specification for sulphate-resisting Portland cement*. BSI, London.
BS 5224: 1976 *Specification for masonry cement*. BSI, London.
BS 1370: 1979 *Specification for low heat Portland cement*. BSI, London.
BS 890: 1972 *Building limes*. BSI, London.
BS 1199 and 1200: 1976 *Building sands from natural sources*. BSI, London.
BS 882: 1983 *Aggregates from natural sources for concrete*. BSI, London.
BS 4887: 1973 *Mortar plastizers*. BSI, London.
BS 1014: 1975 *Pigments for Portland cement and Portland cement products*. BSI, London.
BS 3148: 1980 *Methods of test for water for making concrete (including notes on the suitability of the water)*.
BS 5628: Part 1: 1978 *Structural use of masonry – unreinforced masonry*. BSI, London.
BS 5628: Part 2: 1985 *Use of masonry – Structural use of reinforced and prestressed masonry*. BSI, London.
BS 4721: 1981 *Ready-mixed building mortars*. BSI, London.
BS 5390: 1976 *Stone masonry*. BSI, London.

BS 6270: Part 1: 1985 *Cleaning and surface repair of buildings – natural stone, cast stone and clay and calcium brick masonry.* BSI, London.

BS 4551: 1980 *Methods of testing mortars, screeds and plasters.* BSI, London.

PD 6472: 1974 *Guide to specifying the quality of building mortars.* BSI, London.

Powers, T. C. (1945) A working hypothesis for further studies of frost resistance of concrete. *A.C.I. Journal Proc.*, Vol. 41, pp. 245–72.

Powers, T. C. (1949) The air entrainment of frost-resistant concrete. *Proc. Highway Res. Board*, Vol. 29, Washington DC, pp 184–202.

Powers, T. C. and Helmutt, R. A. (1953). Theory of volume changes in hardened Portland cement paste during freezing. *Proc. Highway Res. Board*, Vol. 32, Washington DC, pp 285–97.

Thomas, K., Coutie, M. G. and Pateman, J. (1970) The effect of pigment on some properties of mortar for brickwork. *SIBMAC Proc. Stoke-on-Trent.*

3
Selection of materials and dimensions

3.1 Selection of masonry units and mortar for durability

The following guidance on the choice of masonry units and mortar designations is based on durability requirements, and other factors such as aesthetics and structural strength may need additional consideration.

If designers are not familiar with the units they propose to use it is strongly recommended that they consult the manufacturer concerned.

The mortar designations are as follows:

Designations	Cement:lime:sand	Cement:sand with plasticizer	Masonry cement: sand
(i)	1:0 to $\frac{1}{4}$:3	–	–
(ii)	1:$\frac{1}{2}$:4 to $4\frac{1}{2}$	1:3 to 4	1:$2\frac{1}{2}$ to $3\frac{1}{2}$
(iii)	1:1:5 to 6	1:5 to 6	1:4 to 5
(iv)	1:2:8 to 9	1:7 to 8	1:$5\frac{1}{2}$ to $6\frac{1}{2}$

Clay brick durability designations are as follows:
FL – frost-resistant (F) and low soluble salt content (L)
FN – frost-resistant (F) and normal soluble salt content (N)
ML – moderately frost-resistant (M) and low soluble salt content (L)
MN – moderately frost-resistant (M) and normal soluble salt content(N)
OL – not frost-resistant (O) and low soluble content (L)
ON – not frost-resistant (O) and normal soluble salt content (N)

Work below or near external ground level

Low risk of saturation with or without freezing

	Units	Mortar designation
Clay bricks	FL, FN, ML or MN	(i) (ii) or (iii)
Calcium silicate bricks	Classes 3 to 7	(iii) or (iv)
Concrete bricks	Compressive strength $\geqslant$ 15 N/mm^2	(iii)
Concrete blocks	(a) Density $\geqslant$ 1500 kg/m^3	(iii)
Concrete blocks	(b) Dense aggregate unit	(iii)
Concrete blocks	(c) Compressive strength $\geqslant$ 7 N/mm^2	(iii)
Concrete blocks	(d) Some AAC blocks	(iii)

High risk of saturation without freezing

	Units	Mortar Designation
Clay bricks	FL or ML (FN or MN)*	(i) or (ii)
Calcium silicate bricks	Classes 3 to 7	(ii) or (iii)
Concrete bricks	Compressive strength $\geqslant$ 15 N/mm^2	(ii) or (iii)
Concrete blocks	(a) Density $\geqslant$ 1500 kg/m^3	(iii)
Concrete blocks	(b) Dense aggregate unit	(iii)
Concrete blocks	(c) Compressive strength $\geqslant$ 7 N/mm^2	(iii)
Concrete blocks	(d) Some AAC blocks	(iii)

High risk of saturation with freezing

	Units	Mortar designation
Clay bricks	FL	(i) or (ii)
Clay bricks	(FN)*	(i) or (ii)
Calcium silicate bricks	Classes 3 to 7	(ii)
Concrete bricks	Compressive strength $\geqslant$ 20 N/mm^2	(ii) or (iii)
Concrete blocks	(a) Density $\geqslant$ 1500 kg/m^3	(iii)
Concrete blocks	(b) Dense aggregate unit	(iii)
Concrete blocks	(c) Compressive strength $\geqslant$ 7 N/mm^2	(iii)
Concrete blocks	(d) Some AAC blocks	(iii)

Some types of autoclaved aerated concrete blocks may not be suitable.
* Sulphate-resisting Portland cement is recommended in this situation.

Where designation (iv) mortar is used it is essential to ensure that all masonry units, mortar and masonry under construction are protected fully from saturation and freezing.

The masonry most vulnerable in the high risk situations with or without freezing is below d.p.c. level and generally within 150 mm above and 150 mm below finished ground level. Masonry within this area will become wet and may remain wet for long periods of time, particularly in winter.

3.2 Damp proof course bricks

In buildings

	Units	Mortar designation
Clay bricks	Damp proof course (1) quality ≤ 4.5 per cent water absorption	(i)

In external works

	Units	Mortar designation
Clay bricks	Damp proof course (2) quality ≤ 7 per cent water absorption	(i)

Brick damp proof courses can resist rising damp but will not resist water percolating downwards. If sulphate ground conditions exist, sulphate-resisting Portland cement is recommended.

Damp proof courses of clay bricks should not be used in conjunction with other masonry units as differential movement may occur.

3.3 Unrendered external walls

This includes walls other than chimneys, cappings, copings, parapets and sills.

Low risk of saturation

	Units	Mortar designation
Clay bricks	FL,FN, ML or MN	(i) (ii) or (iii)
Calcium silicate bricks	Classes 2 to 7	(iii) or (iv)
Concrete bricks	Compressive strength $\geqslant 7$ N/mm^2	(iii)
Concrete blocks	Any type	(iii) or (iv)

High risk of saturation

	Units	Mortar designation
Clay bricks	(FN)* or FL	(i) (ii)
Calcium silicate bricks	Classes 2 to 7	(iii)
Concrete bricks	Compressive strength $\geqslant 15$ N/mm^2	(iii)
Concrete blocks	Any type	(iii)

* Sulphate-resisting Portland cement is recommended in this situation.

Walls should be protected by an overhanging roof or similar projection to offer maximum protection.

Where designation (iv) mortar is used it is essential that all masonry units, mortar and masonry under construction are protected fully from saturation and freezing.

Cappings are defined as an assemblage of units placed at the head of a wall which do not shed rainwater from the top of the wall clear of all exposed surfaces of the walling. Copings are designed to shed rainwater from the top of the wall clear of all exposed faces of the wall they are intended to protect. See Figures 3.1 and 3.2.

3.4   Rendered external walls

This includes walls other than chimneys, cappings, copings, parapets and sills.

	Units	*Mortar designation*
Clay bricks	(FN or MN)*	(i) or (ii)
Clay bricks	FL or ML	(i) (ii) or (iii)
Calcium silicate bricks	Classes 2 to 7	(iii) or (iv)
Concrete bricks	Compressive strength $\geqslant 7 \text{ N/mm}^2$	(iii)
Concrete blocks	Any type	(iii) or (iv)

* Sulphate-resisting Portland cement is recommended in this situation also in the base coat of the render.

Where designation (iv) mortar is used it is essential to ensure that all masonry units, mortar and masonry under construction are fully protected from saturation and freezing.

3.5   Internal walls and inner leaves of cavity walls

	Units	*Mortar designation*
Clay bricks	FL, FN, ML, MN, OL or ON	(i) (ii) (iii) or (iv)
Calcium silicate bricks	Classes 2 to 7	(iii) or (iv)
Concrete bricks	Compressive strength $\geqslant 7 \text{ N/mm}^2$	(iv)
Concrete blocks	Any type	(iii) or (iv)

Where designation (iv) mortar is used it is essential to ensure that all masonry units, mortar and masonry under construction are fully protected from saturation and freezing.

3.6   Unrendered parapets

This excludes other than cappings and copings.

Low risk of saturation
Examples of this catorgory are low parapets on some single storey buildings.

	Units	Mortar designation
Clay bricks	FL, FN, ML or MN	(i) (ii) or (iii)
Calcium silicate bricks	Classes 3 to 7	(iii)
Concrete bricks	Compressive strength $\geqslant 20$ N/mm^2	(iii)
Concrete blocks	(a) Density $\geqslant 1500$ kg/m^3	(iii)
Concrete blocks	(b) Dense aggregate unit	(iii)
Concrete blocks	(c) Compressive strength $\geqslant 7$ N/mm^2	(iii)
Concrete blocks	(d) Most AAC blocks	(iii)

High risk of saturation
Examples are where a capping only is provided for the masonry.

	Units	Mortar designation
Clay bricks	(FN or MN)* or FL or ML	(i) or (ii)
Calcium silicate bricks	Classes 3 to 7	(iii)
Concrete bricks	Compressive strength $\geqslant 20$ N/mm^2	(iii)
Concrete blocks	(a) Density $\geqslant 1500$ kg/m^3	(ii)
Concrete blocks	(b) Dense aggregate unit	(ii)
Concrete blocks	(c) Compressive strength $\geqslant 7$ N/mm^2	(ii)
Concrete blocks	(d) Most AAC blocks	(ii)

* Sulphate-resisting Portland cement is recommended in this situation.

Most parapets are likely to be severely exposed irrespective of the climate exposure of the building. Copings and d.p.c.s should always be provided. Some types of autoclaved aerated concrete blocks may not be suitable for this exposure condition.

3.7 Rendered parapets

This excludes cappings and copings.

	Units	Mortar designation
Clay bricks	(FN or MN)*	(i) or (ii)
Clay bricks	FL or ML	(i) (ii) or (iii)
Calcium silicate bricks	Classes 3 to 7	(iii)
Concrete bricks	Compressive strength $\geqslant 7$ N/mm^2	(iii)
Concrete blocks	Any type	(iii)

* Sulphate-resisting Portland cement is recommended in this situation also in the base coat of the render.

Single leaf walls should only be rendered on one face and all parapets should be provided with a coping.

3.8   Chimneys

Unrendered with low risk of saturation

	Units	Mortar designation*
Clay bricks	FL, FN, ML or MN	(i) (ii) or (iii)
Calcium silicate bricks	Classes 3 to 7	(iii)
Concrete bricks	Compressive strength $\geqslant 10$ N/mm^2	(iii)
Concrete blocks	Any type	(iii)

Unrendered with high risk of saturation

	Units	Mortar designation*
Clay bricks	FL	(i) or (ii)
Clay bricks	FN*	(i) or (ii)
Calcium silicate bricks	Classes 3 to 7	(iii)
Concrete bricks	Compressive strength $\geqslant 15$ N/mm^2	(iii)
Concrete blocks	(a) Density $\geqslant 1500$ kg/m^3	(ii)
Concrete blocks	(b) Dense aggregate unit	(ii)
Concrete blocks	(c) Compressive strength $\geqslant 7$ N/mm^2	(ii)
Concrete blocks	(d) Most AAC blocks$^+$	(ii)

Rendered

	Units	Mortar designation
Clay bricks	(FL or ML)*	(i) (ii) or (iii)
Clay bricks	(FN or MN)*	(i) or (ii)
Calcium silicate bricks	Classes 3 to 7	(iii)
Concrete bricks	Compressive strength $\geqslant 7$ N/mm^2	(iii)
Concrete blocks	Any type	(iii)

* Due to the danger of sulphate attack from flue gases, sulphate-resisting Portland cement is strongly recommended in the mortar and in any rendering.
+ Some types of AAC blocks are not suitable for use in the unrendered, high-risk saturation situation.
Masonry and tile cappings cannot be relied upon to keep out moisture indefinitely. The use of a coping is to be preferred.

Chimney stacks are normally the most exposed masonry on any building.

3.9   Cappings, copings and sills

	Units	Mortar designation
Clay units	FL or FN	(i)
Calcium silicate bricks	Classes 4 to 7	(ii)
Concrete bricks	Compressive strength $\geqslant$ 30 N/mm^2	(ii)
Concrete blocks	(a) Density $\geqslant$ 1500 kg/m^3	(ii)
Concrete blocks	(b) Dense aggregate unit	(ii)
Concrete blocks	(c) Compressive strength $\geqslant$ 7 N/mm^2	(ii)
Concrete blocks	(d) Most AAC blocks$^+$	(ii)

Some autoclaved aerated concrete blocks may be unsuitable for use in the above. When cappings or copings are used for chimney terminals, the use of sulphate-resisting Portland cement is strongly recommended. D.p.c.s for cappings, copings and sills should be bedded in the same grade of mortar as the masonry units.

3.10   Freestanding boundary and screen walls

This excludes cappings and copings.

With coping

	Units	Mortar designation
Clay bricks	(FN or MN)*	(i) or (ii)
Clay bricks	FL or ML	(i) (ii) or (iii)
Calcium silicate bricks	Classes 3 to 7	(iii)
Concrete bricks	Compressive strength $\geqslant$ 15 N/mm^2	(iii)
Concrete blocks	Any type	(iii)

With cappings

	Units	Mortar designation
Clay bricks	FL or (FN)*	(i) or (ii)
Calcium silicate bricks	Classes 3 to 7	(iii)
Concrete bricks	Compressive strength $\geqslant$ 20 N/mm^2	(iii)
Concrete blocks	(a) Density $\geqslant$ 1500 kg/m^3	(ii)
Concrete blocks	(b) Dense aggregate unit	(ii)
Concrete blocks	(c) Compressive strength $\geqslant$ 7 N/mm^2	(ii)
Concrete blocks	(d) Most AAC blocks	(ii)

* Sulphate-resisting Portland cement is strongly recommended.

Most free-standing walls are likely to be severely exposed irrespective of climatic conditions. Such walls should be protected by a coping and d.p.c.s should be provided under the copings and at the base of the wall.

Where designation (iii) mortar is used in the 'with cappings' situation the use of sulphate-resistant Portland cement is strongly recommended. Some types of autoclaved aerated concrete blocks may not be suitable.

3.11   Earth retaining walls

This excludes cappings and copings.

With waterproofed retaining face and coping

	Units	*Mortar designation*
Clay bricks	(FN or MN)*	(i) or (ii)
Clay bricks	FL or ML	(i) or (ii)
Calcium silicate bricks	Classes 3 to 7	(ii) or (iii)
Concrete bricks	Compressive strength $\geqslant$ 15 N/mm^2	(ii)
Concrete blocks	(a) Density $\geqslant$ 1500 kg/m^3	(ii)
Concrete blocks	(b) Dense aggregate unit	(ii)
Concrete blocks	(c) Compressive strength $\geqslant$ 7 N/mm^2	(ii)
Concrete blocks	(d) Most AAC blocks	(ii)

With copings or cappings but no waterproofing on retaining face

	Units	*Mortar designation*
Clay bricks	(FN)*	(i)
Clay bricks	FL	(i)
Calcium silicate bricks	Classes 4 to 7	(ii)
Concrete bricks	Compressive strength $\geqslant$ 30 N/mm^2	(i) or (ii)
Concrete blocks	Certain blocks	(i) or (ii)

* Sulphate-resisting Portland cement is strongly recommended.

Because of the danger of contamination from the ground and saturation by ground waters, in addition to possible severe climatic exposure, masonry in retaining walls is particularly prone to frost and sulphate attack. It is therefore vital that considerable care be taken to select the correct materials and to ensure the exclusion of water.

It is strongly recommended that such walls be backfilled with free-draining material and that the retaining face of the wall be waterproofed. It is also recommended that an effective coping with a d.p.c. be provided.

Some types of autoclaved aerated concrete blocks are not suitable for use with waterproofed retaining face and coping. Most concrete blocks are not suitable for use with coping without waterproofing in the retaining face. Manufacturers should be consulted.

3.12   Drainage and sewerage

Examples are inspection chambers and manholes.

Surface water

	Units	Mortar designation
Clay bricks	Engineering, FL or ML	(i)
	(FN or MN)*	(i)
Calcium silicate bricks	Classes 3 to 7	(ii) or (iii)
Concrete bricks	Compressive strength ≥ 20 N/mm²	(iii)
Concrete blocks	(a) Density ≥ 1500 kg/m³	(ii)
Concrete blocks	(b) Dense aggregate blocks	(ii)
Concrete blocks	(c) Compressive strength ≥ 7 N/mm²	(iii)
Concrete blocks	(d) Most AAC blocks	(ii)

Foul drainage (continuous contact with masonry)

	Units	Mortar designation
Clay bricks	Engineering, FL or ML	(i)
	(FN or MN)*	(i)
Calcium silicate bricks	Class 7	(ii)
Concrete bricks	Compressive strength ≥ 40 N/mm² with cement content ≥ 350 kg/m³	(i) or (ii)
Concrete blocks	Not suitable	

Foul drainage (occasional contact with masonry)

	Units	Mortar designation
Clay bricks	Engineering, FL or ML	(i)
	(FN or MN)*	(i)
Calcium silicate bricks	Classes 3 to 7	(ii) or (iii)
Concrete bricks	Compressive strength ≥ 40 N/mm² with cement content ≥ 350 kg/m³	(i) or (ii)
Concrete blocks	Not suitable	

* Sulphate-resisting Portland cement should be specified.

If sulphate ground conditions exist special precautions need to be taken – see Clause 22.4 of BS 5628: Part 3: 1985.

Some types of autoclaved aerated concrete blocks are not suitable for use in suface water situations. Also some types of calcium silicate bricks are not suitable for use in foul drainage situations (continuous or occasional contact with masonry).

3.13   Damp proof course (d.p.c.) materials

Damp proof courses have been included in masonry structures since the end of the last century to prevent the migration of water from one part of the building to another. Rising dampness from wet areas such as foundations can effectively be stopped by such water barriers. In addition, at a higher level when parapet walls or chimneys form part of a building, the downward passage of rainwater can be diverted, thus eliminating the many problems which can result due to dampness.

Materials

According to BS 5628: Part 3: 1985 and BS 743: 1970 materials for damp proof courses should comply with the following British Standards:

BS 6398: 1983 *Bitumen damp proof courses for masonry*
These d.p.c.s are supplied in rolls of various widths and usually consist of hessian; fibre; asbestos; hessian laminated with lead; fibre laminated with lead or asbestos laminated with lead between two layers of bitumen stabilized by mineral filler and finished with a surfacing material.

When laid they should be lapped at least 100 mm and need to be welted if used to prevent the migration of water in a downwards direction. Bitumen-type d.p.c.s are likely to extrude under heavy loading, particularly in south-facing elevations but this is unlikely to affect their resistance to water penetration. In the extreme it may be responsible for cracking of the masonry.

BS 3921: 1985 *Clay bricks*
Bricks for d.p.c.s are required to have an average absorption by weight of not more than 4.5 per cent for Class 1 and 7 per cent for Class 2. To form an effective d.p.c. a minimum of two courses of the above bricks need to be laid in break joint and bonded in a mortar designation (i), i.e. 1:3 Portland cement:sand mortar. Damp proof course Class 1 should be used for buildings and Class 2 may be used for some categories of external works.

Brick d.p.c.s have the advantage of offering the same tensile strength as the remainder of the wall but they are not suitable to resist the downward passage of water. They have excellent durability and are ideal for use in heavily stressed clay masonry. They should never be used in conjunction with concrete or calcium silicate units.

BS 6515: 1984 *Polythene damp proof courses for masonry*
These d.p.c.s are supplied in rolls in various widths and consist of polythene sheet, incorporating carbon black pigment to protect the material from the effects of ultra-violet light. The sheet material is required to be not less than 0.46 mm thickness (weighing approximately 0.48 kg/m^2) and when laid should be lapped a minimum of 100 mm or the width of the d.p.c. whichever is the greater. Welted joints must be held in compression and when used for cavity trays may need bedding in mastic in the outer leaf to prevent penetration by driving rain. This material does not extrude under normal loading and there is no evidence of deterioration when in contact with other building materials. It is tough and resilient and remains flexible over the temperature ranges likely to occur in service when used in an orthodox manner. At temperatures below 0°C it is vulnerable to impact damage.

Lead
Lead for damp proof courses is required to comply with BS 1178: 1969 and weigh not less than 19.5 kg/m^2 (Code No. 4 : 1.80 mm thickness).

Lead can corrode when in contact with mortar and should therefore be protected on both sides by a coating of bitumen or bitumen paint. It is also wise to coat the mortar bed with bitumen to make sure that there is no contact between the lead and the mortar.

When laid, these d.p.c.s should be lapped at least 100 mm (preferably 150 mm) and need to be welted if used to prevent the migration of water in a downwards direction. These d.p.c.s do not extrude under normal constructional loading but are not generally recommended for high rise structural masonry.

Copper
Copper for damp proof courses is required to comply with BS 2870: 1980 and should be annealed (Grade A).

D.p.c.s which project from the face of the wall to form a drip are required to be formed either in one piece not thinner than 0.46 mm – weight 4.12 kg/m^2 or in two pieces, the d.p.c. being a minimum of 0.25 mm thickness (2.28 kg/m^2 weight) and the projecting drip or flashing 0.50 mm (4.56 kg/m^2). The drip or flashing is required to have a minimum embedment of 50 mm.

When laid, these d.p.c.s should be lapped at least 100 mm (preferably 150 mm) and need to be welted if used to prevent the migration of water in a downward direction. Lapped joints are improved if sealed with bitumen. Water draining from copper d.p.c.s and flashings can stain masonry. After possible initial corrosion when in contact with wet mortar they are usually satisfactory for d.p.c. purposes. However, when soluble salts are known to be present (e.g. from some clay bricks, sea salt in the sand, etc.) a similar treatment to lead is recommended.

Mastic asphalt
Mastic asphalt for damp proof courses is required to conform with either BS 1097: 1973 or BS 1418: 1973.

This is the only truly jointless d.p.c. but is liable to extrude under pressures above 65 kN/m^2 particularly on south-facing elevations. The mortar bed on which it is laid should be scored and brushed free of dust to provide a good key. To ensure a key for the mortar, grit should be beaten into the asphalt immediately after application and left proud of the surface. Alternatively the surface should be scored while still warm.

This type of d.p.c. has excellent durability but is not recommended for structural masonry.

Pitch/polymer
Pitch/polymer for damp proof courses is required to conform with the appropriate clauses of BS 2782: 1978. Some of these d.p.c.s have Agrément certificates.

They are made from pitch, PVC, fillers and plasticizers, reinforced with synthetic fibres. When laid these d.p.c.s should be lapped at least 100 mm and if subjected to the downward passage of water all joints should be sealed.

These d.p.c.s must be unrolled carefully on to the mortar bed, particularly in cold weather, to avoid cracking. At temperatures below 0°C care should be taken to avoid heavy impact; cracking may also be caused by bending the material over sharp edges. Pitch/polymer d.p.c.s have excellent durability and are suitable for use with structural masonry.

Slate
Slate for damp proof courses should be not less than 230 mm long and have a minimum thickness of 4.0 mm. The slate should satisfy the appropriate clauses of BS 5642 : Part 2: 1983. A slate d.p.c. is required to consist of at least two courses of slates, laid to break joint, each slate being bedded in 1:3 Portland cement:sand mortar designation (i).

Epoxy resin/sand mortar
A mixture of epoxy resin and sand is sometimes used as a d.p.c. but this is not included in the list of materials in BS 743: 1970 or BS 5628 : Part 3: 1985.

This mixture forms a rigid d.p.c. with good adhesion. As there are no jointing problems it can be used to resist the upward and downward passage of water. The resin content should be about 15 per cent and the appropriate grade of hardener used following the manufacturer's instructions. The material does not extrude under load and there is no evidence of deterioration when in contact with other building materials.

Method of specifying d.p.c.s

(1) Specify the material, in the case of bitumen, type letter.
(2) Specify the British Standard and/or the appropriate clauses.
     Alternatively the Agrément Certificate (number and title).
(3) State the width, number of rolls and length also the thickness and density.

Precautions

(1) Select the most appropriate d.p.c. for the job – the cheapest may prove false economy.
(2) Unroll bitumen/pitch polymer materials carefully, especially in cold weather.
(3) Make sure joints in flexible d.p.c.s are satisfactory.
(4) Ensure good bonding between slates or bricks and mortar.
(5) Do not use clay brick d.p.c.s with concrete or calcium silicate masonry.
(6) Bed the d.p.c. within a mortar joint.
(7) Most d.p.c. materials produce a plane of weakness in the wall and this should be taken into account when considering the stability of the masonry.

3.14 Wall ties

The functions of a wall tie vary depending upon the type of construction and the loading conditions and the specifier should carefully select a tie of the most appropriate stiffness. The number of ties used can always be varied to meet differing loading requirements but the following criteria need to be considered before the tie is selected.

(1) The ability to transmit the anticipated tensile and compressive forces without excessive deformation.
(2) The ability to transmit shear forces (if required).
(3) The ability to allow vertical differential movement of the two leaves (particularly where the inner leaf is of timber-framed construction).
(4) The ability to allow horizontal differential movement of the two leaves.
(5) The ability to perform satisfactorily during a fire.
(6) The ability to resist corrosion or other forms of degradation.
(7) The ability to resist the passage of water from one leaf to the other.
(8) The ability to offer as small a horizontal area as possible in the centre section to minimize the retention of mortar droppings.
(9) Flexibility to allow for inaccuracies in coursing.
(10) The ability to allow the addition of a device for holding back insulation bats where necessary.

Variety

A wide variety of wall ties is available but **BS 1243 : 1978** '*Metal ties for cavity wall construction*' only includes three types:

(a) The wire 'butterfly' tie (Figure 3.1).
(b) The wire 'double triangle' tie (Figure 3.2).
(c) The strip metal 'vertical-twist' tie (Figure 3.3).

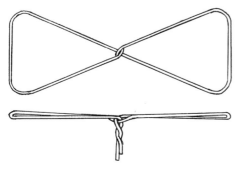

Figure 3.1   Butterfly-type wall tie

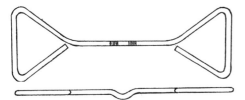

Figure 3.2   Double triangle-type wall tie

Figure 3.3   Vertical twist-type wall tie

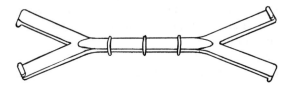

Figure 3.4   Polypropylene wall tie

Polypropylene wall ties are also in common use (Figure 3.4).
   The following is a selection of 'special' ties currently available:

(a)   The Abbey slot-type tie (for use with concrete framed construction) (Figure 3.5).
(b)   Wide cavity/insulation tie (Figure 3.6).
(c)   Chevron (strip steel) tie (for use with timber framed construction) (Figure 3.7).

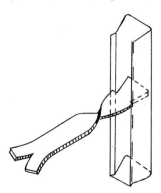

Figure 3.5   Patent Abbey anchor-type wall tie

Materials

BS 1243 : 1978 requires wall ties to be manufactured from one of the following materials:

(1)   Mild steel wire or strip with a zinc coating of not less than 940 g/m^2.
(2)   Plastic coated wire with a zinc coating of not less than 260 g/m^2.

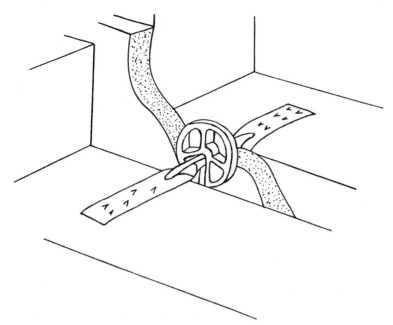

Figure 3.6   Wide cavity/insulation-type wall tie

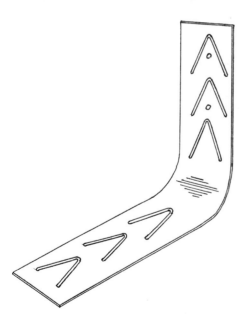

Figure 3.7   Chevron (strip steel)-type wall tie used for timber framed construction

(3)   Copper having a minimum copper content of 99.85 per cent or copper alloys (56 per cent) except for phosphor bronze complying with BS 2870: 1980, BS 2873: 1969 or BS 2874: 1969.

(4)   Austenitic stainless steel wire (BS 1554: 1981) or Austenitic stainless-steel strip (BS 1449 : Part 2: 1970 and BS 970 : Part 4: 1970) minimum 18/8 composition.

NHBC accept austenitic stainless steel, phosphor bronze, silicon bronze or a tie which has a British Board of Agrément Certificate. It is vitally important that wall ties and any fixings they might have are compatible with each other (i.e. dissimilar metals should not be in close contact if electrolytic corrosion is to be avoided).

If ties are to be used in black ash mortar or with marine aggregates (this is not recommended), additional precautions should be taken, such as coating with bitumen, preferably on site.

Placing of ties

The leaves of a cavity should be tied together by wall ties embedded in the horizontal mortar joints at the time the course is laid, to a minimum depth of 50 mm. The length of the wall tie should be chosen to suit the width between the two leaves and laid perpendicular to the leaves it unites. Ties should be placed at a frequency not less than shown in Table 3.1. Additional ties should be provided within 225 mm of all openings so that there is one for each 300 mm of height of the opening. Additional ties will also be necessary adjacent to movement joints.

TABLE 3.1

(A)   Spacing of wall ties

Least leaf thickness (one or both)	Type of tie	Cavity width	Equivalent no. of wall ties per m^2	Horizontal spacing	Vertical spacing
65–90 mm	All	50–75 mm	4.9	450 mm	450 mm
90 mm or more	See Table (B)	50–150mm	2.5	900 mm	450 mm
90 mm or more	Flexible for timber frame	50 mm	4.4	600 mm*	375 mm

(B)   Selection of wall ties

			Type of wall tie		Cavity width
▲ Increasing strength	Increasing flexibility and sound insulation ▼		Vertical twist		150 mm or less
			Double triangle		75 mm or less
			Butterfly		75 mm or less
			Chevron or similar+		50 mm

* This dimension is normally 600 mm but could be less to suit closer spacing of the timber studs.
+Flexible wall ties designed especially for timber framed construction.

Method of specifying wall ties

(i) When appropriate, specify that the wall ties will comply with the appropriate clauses of BS 1243 : 1978 '*Metal ties for cavity wall construction*'. Alternatively, that the wall ties will comply with the appropriate British Board of Agrément Certificate (quoting the certificate title and number).

(ii) State (a) The quantity required
        (b) The type of material and minimum tensile strength
        (c) The dimensions
        (d) Any special marking requirements
        (e) Any test data required.

Precautions

(i)   Beware of wall ties not conforming with BS 1243 : 1978 or not having a BBA certificate. Substandard wall ties are available.

(ii)   Use suitable wall ties for the project.

(iii) Do not use mortar having corrosive properties (e.g. mortars containing frost inhibitors based on calcium chloride, black ash or unwashed marine aggregates).

(iv)   Do not push wall ties into mortar joints after they are made or place them in perpend joints.

(v)   Do not slope wall ties down towards the inner leaf.

(vi)   Do not use stiff wall ties in separating walls, otherwise the sound insulation will be reduced.

3.15   Brickwork dimensions

Bricks are available in a range of sizes but the most commonly used brick has a work size of 215 mm length × 102.5 mm width × 65 mm height. For the purpose of the following examples only Standard size bricks will be considered.

Dimensional tolerances vary with the type of brick used, the material and the method of manufacture and the appropriate British Standards should be consulted for actual values. When brickwork dimensions are being considered it is preferable to work in standard unit sizes to avoid cutting and unsightly jointing.

Overall dimensions of brickwork using standard bricks are normally calculated using 10 mm joints thus the format or co-ordinating size becomes 215 + 10 mm = 225, 102.5 (no joint for single leaf wall) = 102.5 and 65 + 10 = 75 mm.

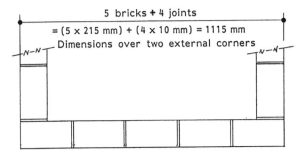

Figure 3.8   Dimensions – 5 bricks + 4 joints

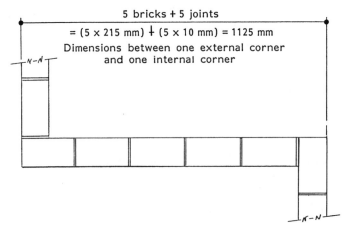

Figure 3.9   Dimensions – 5 bricks + 5 joints

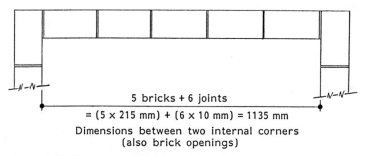

Figure 3.10   Dimensions – 5 bricks + 6 joints

Example – horizontal brickwork dimensions

Calculating panel width – Standard bricks with 10 mm joints (Table 3.2). Movement joints will be required in long lengths of brickwork. If 10 mm movement joints are used these will normally be at centres not exceeding 12 m for clay brickwork; 9 m for calcium silicate brickwork and 6 m for concrete – see Chapter 4 for further details.

Non-standard dimensions

When non-standard dimensions are used this can cause problems if cutting of the units is to be avoided. In long lengths of brickwork it may be possible to avoid cutting of bricks by varying the joint sizes but in smaller panels it will be necessary to introduce cut bricks. If cut bricks are unavoidable it is preferable to locate these adjacent to internal corners where they are less noticeable. Cut bricks introduced at mid-panel tend to be unsightly, similarly at prominent external corners.

Brick piers

Problems may be encountered in brick piers and short lengths of brickwork if variation in brick sizes are outside normal tolerances. In some instances bricks

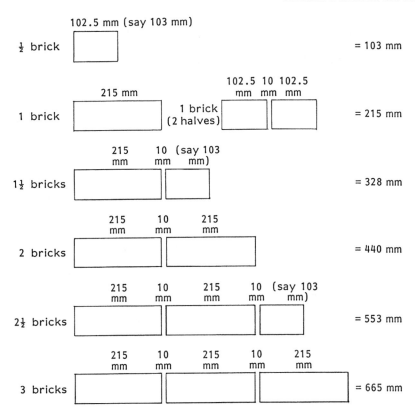

Figure 3.11   Ready reckoner for standard brickwork – horizontal dimensions

TABLE 3.2 Horizontal brickwork dimensions

		Number of bricks /length (mm)				
$\frac{1}{2}$ brick	103	$10\frac{1}{2}$ brick	2353	$20\frac{1}{2}$ brick	4603
1	215	11	2465	21	4715
$1\frac{1}{2}$	328	$11\frac{1}{2}$	2578	$21\frac{1}{2}$	4828
2	440	12	2690	22	4940
$2\frac{1}{2}$	553	$12\frac{1}{2}$	2803	$22\frac{1}{2}$	5053
3	665	13	2915	23	5165
$3\frac{1}{2}$	778	$13\frac{1}{2}$	3028	$23\frac{1}{2}$	5278
4	890	14	3140	24	5390
$4\frac{1}{2}$	1003	$14\frac{1}{2}$	3253	$24\frac{1}{2}$	5503
5	1115	15	3365	25	5616
$5\frac{1}{2}$	1228	$15\frac{1}{2}$	3478	$25\frac{1}{2}$	5728
6	1340	16	3590	26	5840
$6\frac{1}{2}$	1453	$16\frac{1}{2}$	3703	$26\frac{1}{2}$	5953
7	1565	17	3815	27	6065
$7\frac{1}{2}$	1678	$17\frac{1}{2}$	3928	$27\frac{1}{2}$	6178
8	1790	18	4040	28	6290
$8\frac{1}{2}$	1903	$18\frac{1}{2}$	4153	$28\frac{1}{2}$	6403
9	2015	19	4265	29	6515
$9\frac{1}{2}$	2128	$19\frac{1}{2}$	4378	$29\frac{1}{2}$	6628
10	2240	20	4490	30	6740

may need to be specially selected (even though they come within the specific limits of the appropriate British Standard) and/or the mortar joint thicknesses will need to be modified to give greater flexibility.

Vertical brickwork dimensions

Vertical brickwork dimensions can be affected by the introduction of d.p.c.s and cavity trays, etc. (Table 3.3). Indeed if the d.p.c. near ground level is correctly introduced into the mortar joint and not (as is common practice) under the joint, it will be necessary to provide a thicker joint than 10 mm – say 16 mm.

It is also important to specify the correct spacing of wall ties which can be achieved with standard 65 mm high bricks and 215 mm blocks – frequently the spacings specified are impossible to achieve.

TABLE 3.3 Vertical brickwork courses dimensions using 65 mm bricks and 10 mm joints

Brick courses	Height (mm)	Block courses	Brick courses	Height (mm)	Block courses
1	075		31	2325	
2	150		32	2400	
3	225	225	33	2475	2475
4	300		34	2550	
5	375		35	2625	
6	450	450	36	2700	2700
7	525		37	2775	
8	600		38	2850	
9	675	675	39	2925	2925
10	750		40	3000	
11	825		41	3075	
12	900	900	42	3150	3150
13	975		43	3225	
14	1050		44	3300	
15	1125	1125	45	3375	3375
16	1200		46	3450	
17	1275		47	3525	
18	1350	1350	48	3600	3600
19	1425		49	3675	
20	1500		50	3750	
21	1575	1575	51	3825	3825
22	1650		52	3900	
23	1725		53	3975	
24	1800	1800	54	4050	4050
25	1875		55	4125	
26	1950		56	4200	
27	2025	2025	57	4275	4275
28	2100		58	4350	
29	2175		59	4425	
30	2250	2250	60	4500	4500

Metric/Imperial brickwork

When metric brickwork is used for extending existing Imperial brickwork problems can occur as the standard format (75 mm) is less than the Imperial dimension of 3 inches (i.e. $2\frac{5}{8}$ in. brick + $\frac{3}{8}$ in. joint) or 76.2 mm. Although the difference in dimen-

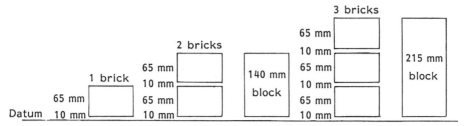

Figure 3.12 Ready reckoner for standard brickwork – vertical dimensions and blockwork.
(**Note**: the illustration does not include a d.p.c.)

sion is small the effect is cumulative and if brick courses in the new work are to line up with the existing, the bed joints will need to be increased in thickness. Alternatively, the new and existing work can be separated visually by a special feature or rainwater downpipe.

Horizontal movement joints

It is essential to make allowance for any differential movement which may occur between the various materials of construction if damage to the structure is to be avoided. This is discussed in detail in Chapter 4. Horizontal movement joints (soft joints) may need to be of a greater thickness than 10 mm and suitable allowance for such joints should be made when considering dimensional tolerances of the brick-work.

Allowance must be made for movements in a vertical direction to accommodate:

(a)   Thermal expansion and contraction of the masonry
(b)   Permanent and cyclic moisture movement of the masonry
(c)   Accidental transfer of load from one leaf of masonry to the other by rigid wall ties
(d)   Drying shrinkage of a reinforced concrete frame
(e)   Elastic and plastic shortening of a reinforced concrete frame
(f)   Deflection and creep of reinforced concrete beams

Dimensions of openings

It is important to coordinate the dimensions of window and door openings in brick walls with the brickwork dimensions. It is all too easy to consider the plan dimensions and to forget the bonding for 'hole in the wall' window openings. Wherever possible windows should be incorporated in the brickwork without the introduction of cut bricks. Account should also be taken of extra joints necessary where steel angles or pressed metal lintels are used for supporting the brickwork. It is bad practice to use the window frames alone for supporting the brickwork unless they are specifically designed for the purpose.

3.16   Blockwork dimensions

The general principles relating to blockwork dimensions follow the same rules as for brickwork. However, BS 6073: Part 2 : 1981 specifies three standard lengths for

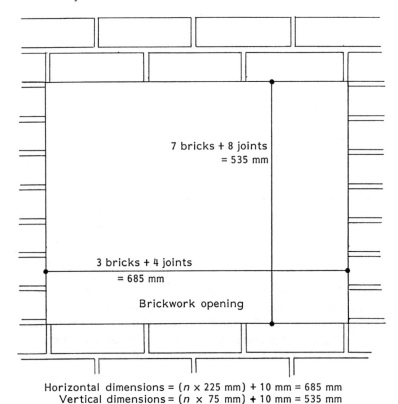

Horizontal dimensions = (n × 225 mm) + 10 mm = 685 mm
Vertical dimensions = (n × 75 mm) + 10 mm = 535 mm

Figure 3.13   Ready reckoner for standard brickwork – brickwork opening. (n = the number of courses of bricks.)

blocks, i.e. work sizes of 390, 440 and 590 mm to provide format or co-ordinating face dimensions of 400, 450 and 600 mm, respectively. The height of the 390 mm block is 190 mm to produce a modular co-ordinating face size of 400 × 200 mm and the other two standard blocks have heights compatible with brick dimensions for use in brick/block cavity wall construction. Additionally the 590 mm block is available in a height of 190 mm for use with the smaller modular blocks. In addition to the standard blocks specified in BS 6073: Part 2: 1981 some manufacturers supply additional work sizes while others do not necessarily produce the complete range.

Various block thicknesses are available as standard but again no single manufacturer necessarily produces the complete range – those listed in the Standard are 60, 75, 90, 100, 115, 125, 140, 150, 175, 190, 200, 215, 220, 225 and 250 mm.

Brick/block cavity construction

The principle of calculating blockwork dimensions is the same regardless of whether this is based on traditional sizes or metric modular sizes. Traditional sizes have been used in this example as it represents the most usual method of construction.

Overall dimensions of blockwork using standard blocks are normally calculated using 10-mm joints and although this is not critical when considering horizontal dimensions for plastered blockwork, it is important to ensure that the bedding joints are in line with the brickwork joints to facilitate wall ties and cavity trays, etc.

Horizontal dimensions

Although any badly cut blocks can be disguised in plastered blockwork, neat work always pays off and it cannot be emphasized too strongly that careful planning aids efficient blocklaying. Cut blocks at corners and at the ends of lintels, etc. are easily incorporated into the work but clay bricks (which have opposing movement characteristics) should never be used as make-up pieces. Compatible concrete bricks can of course be used where appropriate. Horizontal dimensions in brick/block cavity construction are dependent upon the cavity widths and Figure 3.14 and Table 3.4 relate to whole block lengths only.

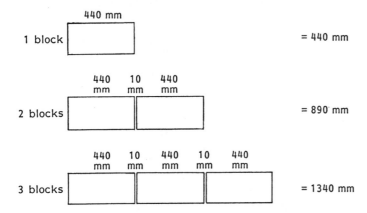

Figure 3.14 Ready reckoner for standard blockwork – horizontal dimensions

TABLE 3.4 Horizontal blockwork dimensions

		Number of blocks/length (mm)			
1	Block	440	—	2	Bricks
2		890	—	4	
3		1340	—	6	
4		1790	—	8	
5		2240	—	10	
6		2690	—	12	
7		3140	—	14	
8		3590	—	16	
9		4040	—	18	
10		4490	—	20	
11		4940	—	22	
12		5390	—	24	
13		5840	—	26	
14		6290	—	28	
15		6740	—	30	
16		7190	—	32	

TABLE 3.5 Vertical blockwork courses dimensions with standard 65 mm bricks and 10 mm joints

Block courses	Height (mm)	Brick courses	Block courses	Height (mm)	Brick courses
	075	1		2325	31
	150	2		2400	32
1	225	3	11	2475	33
	300	4		2550	34
	375	5		2625	35
2	450	6	12	2700	36
	525	7		2775	37
	600	8		2850	38
3	675	9	13	2925	39
	750	10		3000	40
	825	11		3075	41
4	900	12	14	3150	42
	975	13		3225	43
	1050	14		3300	44
5	1125	15	15	3375	45
	1200	16		3450	46
	1275	17		3525	47
6	1350	18	16	3600	48
	1425	19		3675	49
	1500	20		3750	50
7	1575	21	17	3825	51
	1650	22		3900	52
	1725	23		3975	53
8	1800	24	18	4050	54
	1875	25		4125	55
	1950	26		4200	56
9	2025	27	19	4275	57
	2100	28		4350	58
	2175	29		4425	59
10	2250	30	20	4500	60

Movement joints will be required in long lengths of blockwork. If 10-mm movement joints are used these will normally be at centres not exceeding 6 m – see Chapter 4 for further details.

Facing blockwork. In facing blockwork the horizontal dimensions are calculated in a similar manner to brick/block cavity construction but 'specials' are sometimes used at corners and reveals.

Window and door openings. The principle of calculating window and door openings is the same for blockwork as for brickwork. However, openings in facing blockwork should generally be storey height using appropriate window and door configurations.

Modular design. Modular design has many advantages and the standard 390 × 190 and 590 × 190 mm blocks are generally produced for modular co-ordinated facing blockwork.

Planning within the size parameters involves careful early consideration which pays dividends not only in the economics it provides but also in the standard of

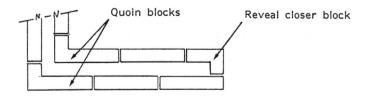

Quoin blocks   Reveal closer block

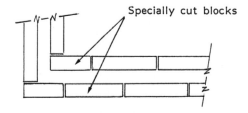

Specially cut blocks

Figure 3.15   Bonding at corners (blockwork)

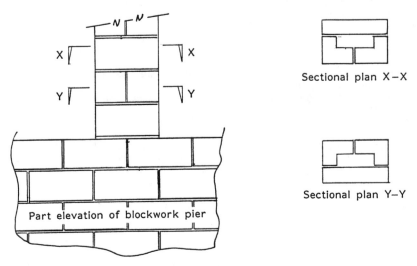

Part elevation of blockwork pier

Sectional plan X–X

Sectional plan Y–Y

Figure 3.16   Blockwork piers. (**Note:** special blocks may be required.)

finish. With fair-faced finishes the most pleasing results are obtained by the use of special reveal closer blocks and quoins.

Special attention should be given to the size of piers, panels between windows and the lengths of walls generally, which should be made in lengths compatible with block modules. Attempts at lengthening the bond by putting in small pieces or widening perpends should be avoided.

The modular principle should be carried right through coping details to avoid the interruption of coursing.

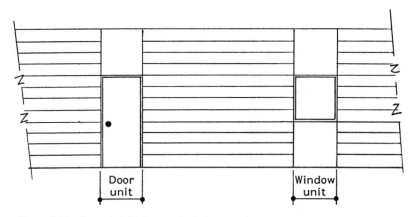

Figure 3.17   Storey height door and window openings

References

BDA Design Note No. 3, July 1979 *Brickwork Dimension Tables.*
BS 5628: Part 3: 1985 *Use of masonry – materials and components, design and workmanship.* BSI, London.
BS 6398: 1983 *Bitumen damp proof courses for masonry.* BSI, London.
BS 3921: 1985 *Clay bricks.* BSI, London.
BS 6515: 1984 *Polythene damp proof courses for masonry.* BSI, London.
BS 743: 1970 *Materials for damp proof courses.* BSI, London.
BS 1178: 1969 *Milled lead sheet and strip for building purposes.* BSI, London.
BS 2870: 1980 *Specification for rolled copper and copper alloys sheet, strip and foil.* BSI, London.
BS 1097: 1973 *Mastic asphalt for building (limestone aggregate).* BSI, London.
BS 1418: 1973 *Mastic asphalt for building (natural rock asphalt aggregate).* BSI, London.
BS 2782: 1978 *Determination of carbon black content.* BSI, London.
BS 5642: Part 2: 1983 *Specification for copings of pre-cast concrete, cast stone, clayward, slate and natural stone.* BSI, London.
BS 1243: 1978 *Metal ties for cavity wall construction.* BSI, London.
BS 2873: 1969 *Copper and copper alloys.* BSI, London.
BS 2874: 1969 *Copper and copper alloys. Rods and sections (other than forging stock).* BSI, London.
BS 1554: 1981 *Specification for stainless and heat-resisting steel round wire.* BSI, London.
BSI 1449: Part 2: 1970 *Stainless and heat resisting steel plate, sheet and strip.* BSI, London.
BS 970: Part 4: 1970 *Stainless, heat resisting and valve steels.* BSI, London.
BS 3921: 1985 *Clay Bricks.* BSI, London.
BS 187: 1978 *Calcium silicate (sandlime and flintlime) bricks.* BSI, London.
BS 6073: Part 2: 1981 *Pre-cast concrete masonry units Specification for pre-cast concrete masonry units.* BSI, London.
BS 6073: Part 2 ... 1981 *Pre-cast concrete masonry units. Method of specifying pre-cast concrete masonry units.* BSI, London.
de Vekey, R. C. 1984 *Performance specifications for wall ties.* BRE.

4

Dimensional changes in masonry and tolerances

Masonry is a composite material made up of individual units plus sand, cement, lime and/or plasticizer which is held, to some extent, or permitted to slide on a d.p.c. and laid with varying degrees of skill which undoubtedly affect the end product and hence its movement characteristics. Dimensional changes due to variations in temperature or moisture content, chemical or frost action, movements in adjacent structural materials and deflections/creep and settlement can occur and may take place as a result of any one of the phenomena mentioned but are quite often due to some combination.

4.1 Thermal movement

Temperatures used for calculating expansion should be the average within the wall. For solid walls, these may be temperatures at the centre of the wall; in cavity wall construction there may be differential thermal movement between the inner and outer leaves and in such situations provision should be made for maximum thermal movement by considering the average temperature of the outer leaf.

Although the movement of internal walls is likely to be much less than that for external walls one must consider inside temperatures as well as the heat transfer characteristics of the construction between the outside and the inside, particular attention being paid when designing internal rooms around refrigerated areas or boiler houses. Special care should also be taken when designing thin walls with a southerly aspect as in some parts of this country surface temperatures can be as high as 65°C, particularly if the cavity is filled with insulation and a dark-coloured unit has been used. Such temperatures can give rise to a thermal gradient through the wall and in cavity construction, even between the two leaves, which may cause excessive bending in addition to longitudinal expansion. Vertical movements in walls are generally reversible but horizontal movements may only be reversible if the wall does not crack as a result of the expansion or contraction. This may depend upon whether the wall is built on a soft/flexible d.p.c., as the degree of restraint imposed by the d.p.c. appears to be a critical factor, also the wall configuration and loading conditions (Figures 4.1–4.3).

The theoretical reversible free movement due to thermal effects is equal to the temperature range multiplied by the appropriate coefficient of thermal expansion.

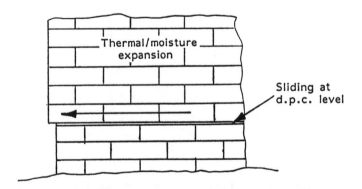

Figure 4.1   Oversailing at d.p.c. level. This is often a sympton of thermal/moisture expansion

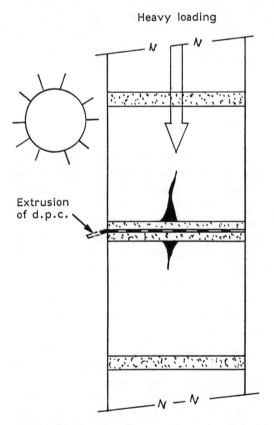

Figure 4.2   Cracking of masonry due to extrusion of d.p.c

Heavy loading

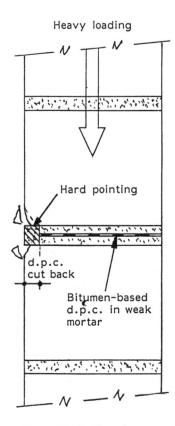

Hard pointing

d.p.c.
cut back

Bitumen-based
d.p.c. in weak
mortar

Figure 4.3 Spalling of masonry due to hard pointing and compressible mortar/d.p.c

However, the movement that actually occurs within a wall after construction depends not only on the range of temperature but also on the initial temperature of the masonry units when laid. This will vary according to the time of year and the conditions during the construction period, as well as on the age of the units when laid and their method of manufacture. Thus in order to determine the potential free movement that could occur in a wall some estimate of the initial temperature and the likely range of temperature should be made.

The potential free movement so deduced will then need to be modified to allow for the effect of restraints. Coefficients of linear thermal expansion are quoted in Chapter 1, Tables 1.10 and 1.12.

4.2 Temperature

BRE Digest 228 (1979) gives examples of service temperature ranges for various materials and the range of temperature quoted for heavyweight walling is 70°C for light colours and 85°C for dark colours. These data are based on a minimum temperature of −20°C and maximum surface temperature of 50°C and 65°C respectively. In the UK, masonry would not be constructed at temperatures as low as

−20°C and Foster and Johnson (1982) have suggested that a reasonable lower bound might be 0°C. This would produce a temperature range 50°C to 65°C. However, taking these figures into account and the research carried out by Beard *et al.* (1969) they suggest an overall range of mean temperature of 45°C at equinoctial periods.

4.3 Moisture movement

The moisture movement of fired-clay masonry units and concrete and calcium silicate masonry units differ in magnitude and in kind. Hence designers should, if they wish to avoid problems, consider the often opposing movement characteristics of the various masonry units and make allowance for such movements in their designs.

More detailed information on moisture movements of the individual units is given in Chapter 1, Tables 1.9 and 1.11.

Mortars

Movement of mortars due to changes in moisture content and carbonation are similar but somewhat greater than for concrete bricks and blocks because of the higher cement content of the mortar, the greater porosity and the fineness of the aggregate.

In practice the movements of the weaker mortars are largely restrained, at least in the plane of the mortar bed, by their adhesion to the bricks and blocks but the movements of the stronger mortars may be sufficient to break this bond.

Typical shrinkage values due to moisture content are 0.04 to 0.10 per cent (initial drying and irreversible) and 0.03 to 0.06 per cent (subsequent and reversible movement). The actual values will depend on the constituents of the mortar, the proportions of the mix and the ambient relative humidity. However, as a general rule, the lower values may be taken to apply to mortars in external walls and the higher values to mortars in internal walls. The resulting movement of internal walls may generally be neglected since they are unlikely to become wet after drying out initially.

4.4 Movement due to chemical action

Sulphate attack and the accompanying mortar expansion is usually confined to clay brickwork or blockwork, although all types of masonry can on occasions suffer when in contact with sulphate-bearing ground or a highly polluted chemical atmosphere. Sulphate attack of clay masonry is most likely to be experienced in unprotected or badly designed parapets, retaining walls and other structures normally liable to remain wet for long periods and then only when bricks or blocks of high soluble sulphate content are used in such situations. It is also important that the units should not be stacked directly on ground which has previously received chemical treatment or alternatively has been used for grazing. Whatever the source of sulphates, expansion of the mortar takes place resulting in movement of the walling (Figure 4.4).

Sulphate expansion can only occur in the presence of three agents, i.e. water-soluble sulphates, tricalcium aluminate (a constituent of Portland cement) and water. Thus, sulphate expansion is unlikely to occur in masonry where the units

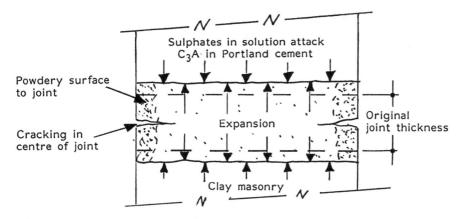

Figure 4.4 Sulphate attack of mortar joints

have negligible sulphate content or the mortar has a low tricalcium aluminate content, as in mortar made from sulphate-resistant cement, or when water is largely excluded by sound methods of building construction. Thus it is easy to visualize conditions in which bricks of moderate salt content could have given good service and other conditions in which bricks of less salt content could have performed badly. There are many other factors too which obscure this issue.

In situations where sulphate expansion is likely to occur it is recommended that sulphate-resistant cement be used (see BRE Digest No. 165, 1974). The movements in the worst cases can be of the order of 2 to 3 per cent, by which time the masonry is in a very unsightly and often unstable condition.

4.5   Movement due to freezing

The movement of masonry due to freezing is a secondary action which can only take place after sulphate or frost attack of the mortar. After the primary failure pores and crevices are opened up in the mortar, thus allowing water penetration and the subsequent expansion upon freezing. It is important therefore to protect masonry during erection to prevent the intrusion of excessive moisture which is liable to cause frost failure in the mortar and expansion in the masonry.

4.6   Effects of associated structure and components

In addition to dimensional changes in the units used for constructing masonry the affects of associated structure, soil/structure interaction and other components need to be considered.

Ground movements

Masonry may be affected by ground movements due to consolidation settlement resulting from construction loading and movements which take place independently of imposed loading, these include:

(i)   Swelling and shrinkage due to varying moisture content, as well as temperature changes under foundations of boiler houses, etc.
(ii)   Frost heave; in the UK this takes place in chalk, chalky soils and silty soils.
(iii) Slopes and landslips.
(iv)   Mining subsidence.
(v)   Shock and vibration.
(vi)   Swallow or sink holes in chalk and limestone areas.

When differential settlement occurs between parts of a building cracking is inevitable. Normal movement joints are of limited value in such situations unless closely spaced, allowing individual panels to rotate thus absorbing movement and preventing cracking.

The avoidance of cracking due to differential settlement is largely a question of foundation engineering with loads distributed according to the bearing capacity of the soil and ground adequately compacted, etc. Where such movement is anticipated in some instances (e.g. when a new building abuts an existing one) it may be advantageous to make a complete break in the form of a temporary joint until differential movement has ceased.

Large expanses of cavity brickwork may be subjected to differential movement due to various causes and Clause 29.2 of BS 5628: Part 1: 1978 'Structural use of masonry' makes the following recommendations:

'The uninterrupted height and length of the outer leaf of external cavity walls should be limited so as to avoid undue loosening of the ties due to differential movements between the two leaves. The outer leaf should, therefore, be supported at intervals not more than every third storey or every 9 m, whichever is less. However, for buildings not exceeding four storeys or 12 m in height, whichever is less, the outer leaf may be uninterrupted for its full height. Consideration should also be given to the provision of vertical joints to accommodate movements (see BS 5628: Part 3: 1985)'.

Differential movement of dissimilar materials

When designing composite walls (i.e. walls involving the use of more than one basic type of unit) it is extremely important to consider the way in which the units behave. Clay masonry for example, expands while concrete and calcium silicate products contract. It is of the utmost importance to realize that the rigid bonding of materials with diametrically opposed movement characteristics can only result in problems, unless the length of wall is extremely short and whenever possible walls faced with units dissimilar from the backing should be of cavity construction using flexible wall ties (Figures 4.1–4.8).

When cavity wall construction is used, it is also important at jambs of openings, etc. where the cavity is closed with masonry and at stop ends to provide a vertical slip plane in the form of a flexible d.p.c.-type material. This is desirable regardless of whether a d.p.c. is necessary to prevent the ingress of moisture.

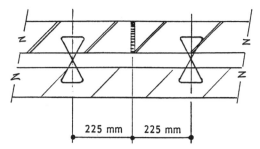

Figure 4.5   Wall ties not more than 225 mm either side of movement joint

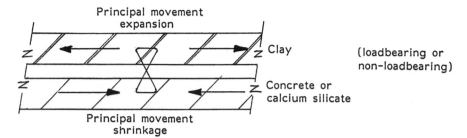

Figure 4.6   Flexible wall ties needed

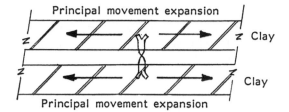

Figure 4.7   Stiff/flexible wall ties. Loadbearing–stiff wall ties required. Non-loadbearing–stiff or flexible wall ties required

Materials

Masonry
The condition of the units when laid can affect the long-term performance of the walls. It is therefore important that they are inspected both when delivered to site and immediately before use, to check whether they have been subject to deterioration or damage. All units should be carefully stacked on pallets or otherwise protected to ensure that they are not contaminated by chemicals or rising moisture. They should also be protected from rain and snow. For concrete and calcium silicate masonry units it is recommended that provision is made for the free circulation of air within the stack so that the units may dry out before building into the work. They should not normally be built into the work for at least 4 weeks at normal temperatures (longer in colder weather); slightly shorter periods of storage may be adequate if the

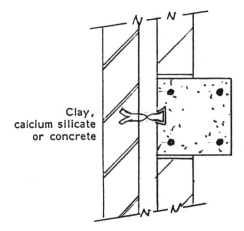

Figure 4.8   Stiff wall ties allowing differential vertical movement

units have been cured in low-pressure steam and cooled so that they have an opportunity of drying. Autoclaved concrete and calcium silicate products need only sufficient storage to allow them to cool. Fired clay units should not be used fresh from
the kiln and a minimum of seven days should be allowed between drawing from the
kiln and laying in the work on site.

Mortar
Although mortar forms only a relatively small proportion of the masonry as a
whole its characteristics nevertheless do have a significant effect particularly in
relation to movements both within the wall itself and with adjacent parts of the
structure.

The traditional lime mortars could accommodate movement from the masonry
units and frequently redistributed the units when ground movements occurred without actual cracking of the masonry. Modern mortars based on ordinary Portland
cement:sand with perhaps additions of lime and/or plasticizer do not have the same
flexibility. The golden rule is never to specify a mortar stronger than necessary for
the job in hand and to use the properties of creep and plastic flow which will
undoubtedly tend to relieve high stresses and reduce the risk of cracking of the
masonry. It is uneconomical and very unwise to specify mortars stronger than
necessary. This is particularly so with low strength units.

Wall ties
The performance of wall ties varies depending upon the type of construction and the
loading and designers should carefully select a tie of the most appropriate stiffness.

If two leaves of clay masonry are used in a loadbearing situation it will be appropriate to use the stiff vertical twist type wall tie. However, if a clay unit is to be used
in conjunction with a lightweight concrete unit then a more flexible type of tie would
generally be recommended as its ability to accommodate a greater degree of differential movement between the two leaves reduces the risk of cracking. In timber-
frame construction the masonry is treated merely as a veneer and wall ties capable

of permitting maximum vertical differential movement (and some horizontal movement) but adequate lateral restraint should be specified. Similarly, when masonry is used to clad reinforced concrete-framed construction differential vertical movement may be critical and special stiff ties such as 'Abbey Anchors' may be appropriate (Figure 4.8).

Cavity wall insulation has also been responsible for a host of innovative wall tie designs which accommodate and support sheets of insulation partially filling the wider than normal cavity and add another factor to be considered when selecting the appropriate wall tie for the project (Figures 4.5–4.8 and figures 4.15 and 4.16).

To summarize, the following factors should be considered when selecting wall ties for cavity wall construction with respect to movement:

(a) Vertical differential movement (Figure 4.8)
(b) Horizontal differential movement (Figures 4.5–4.7)
(c) Lateral movement due to wind forces (pressure and suction)
(d) Racking movement due to wind forces (pressure and suction) (Figure 4.9)
(e) Accidental transfer of loadbearing/non-loadbearing leaves (Figure 4.10).

Damp proof course materials
Damp proof course materials (Figures 4.1–4.3) used in modern wall construction usually consist of a flexible membrane such as bituminous felt, pitch polymer or polyethylene. However, bricks for damp proof courses laid in an appropriate mortar are sometimes used in conjunction with fired-clay masonry.

The type of d.p.c. used has a considerable effect upon the degree of movement permitted in walling, the brick d.p.c. restraining movement to a certain extent and the flexible membrane permitting greater movement and often acting as a break–bond/slip plane.

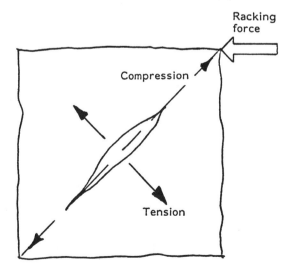

Figure 4.9   Racking force causing tensile splitting

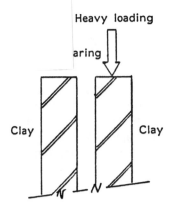

Heavy loading

aring

Clay

Clay

Figure 4.10 Stiff ties – provided it is acceptable to share the bending induced

If flexible d.p.c.s are used they should not be cut back from the face of the brickwork and pointed over to disguise their presence as this can not only make them ineffective but can also encourage differential loading and spalling of the masonry and provide a partial restraint at the outer edge of the wall. Cracking and spalling of walls frequently occurs at corners when flexible d.p.c.s stop short or are incorrectly lapped, thus creating a point of accidental restraint (Figures 4.2 and 4.3).

Special care should be taken when selecting d.p.c.s for highly stressed masonry as it is important that the material should have low creep properties to avoid cracking of the masonry. The choice of d.p.c. and the method of incorporating it into the wall undoubtedly play a major role in determining whether or not the wall will perform in a satisfactory manner. Figures 4.11–14 and Figure 4.17 illustrate some of the problems associated with dimensional changes in masonry.

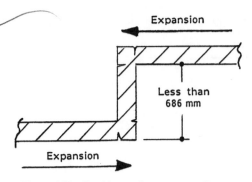

Expansion

Less than
686 mm

Expansion

Figure 4.11 Cracking at short returns. (Larger returns tend to rotate without cracking.)

4.7 Movement joints

In long walls or buildings of complex shape movement joints are too often omitted and while structural failure of masonry is rare, unsightly cracking can usually be avoided if the appropriate measures are taken at the design stage.

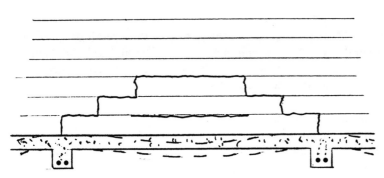

Figure 4.12   Masonry cracking due to excessively deflecting floors. (Reinforcement in lower joints can prevent cracking.)

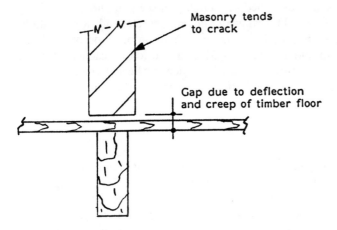

Figure 4.13   Avoid masonry supported on timber floors

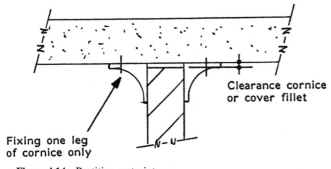

Figure 4.14   Partition restraint

Location of joints

Movement joints should be positioned at centres to suit the overall design of the building and its general aesthetics, bearing in mind the materials used and its orientation. Additional features of the building which should be considered when determining joint positions in masonry are as follows:

(i)   intersecting walls, piers, floors etc
(ii)   window and door openings
(iii) change in height or thickness of the wall
(iv)   chases in the wall
(v)   movement joints in the structural framework or in floor slabs.

Areas above doors and above or below windows may need to be reinforced or alternatively constructed in another material compatible with the construction.

It is vital that when joints are positioned the stability of the structure is in no way impaired. Joints are often concealed by rainwater downpipes, etc. and care should be taken to ensure that fixing of lugs or brackets occur only on one side of the joint. Rendering and plastering should not be carried over the joints, each coat being severed by a well defined cut before the work hardens. Alternatively, rendering can take place with a temporarily secured batten to the edge of the joint or a vertical d.p.c.

The movement joint

The proportions of movement joints and the seal are important. To ensure adequate bond to the masonry the depth of seal should be at least 10 mm. Certain single-part moisture-cured sealants are best used in joints of small cross-section due to their excessive curing time in thick sections. Optimum performance in butt joints is obtained when the width-to-depth ratio of the sealant bead lies within the range 2:1 to 1:1, for elastoplastic sealants (including one and two part polysulphides) or the range 1:1 to 1:2 for plastoelastic sealants (including cross-linked butyl rubber). The sealant should be applied against a firm jointing material in order to force it against the sides of the joint with sufficient pressure to ensure good adhesion. The joint filler should be resilient and not adhere to or react with the sealant. The compressibility of the jointing material is perhaps the most critical factor in the design of the joint for expansion joints (i.e. for fired-clay masonry). According to BS 5628: Part 3: 1985 a pressure of about 0.1 N/mm^2 should be sufficient to compress the material to 50 per cent of its original thickness. Flexible cellular polyethylene, cellular polyurethane or foam rubbers are considered to be the most satisfactory materials. Hemp, fibreboard, cork and similar materials should not be used for expansion joints in fired-clay masonry but may be suitable for contraction or control joints in concrete and calcium silicate masonry.

The width of the joint must be sufficient to accommodate the possible movements, both reversible and irreversible, without damage to the seal. Hence the width of the joints should be related to their centres.

For further guidance on the selection of sealants, designers are referred to the manufacturers and to BS 6213: 1982.

Provision of movement joints

The data for calculating theoretical movements in the various materials given earlier may be considered somewhat complex and is included for designers who wish to investigate the various movements in detail and determine joint widths and centres for buildings which may have unusual features. Alternatively the data may be useful when investigating building failures.

For the more general type of construction the following empirical recommendations may be found applicable to the majority of situations:

(i)   Movement joints are not normally required in internal walls in dwellings.
(ii)   The spacing of the first movement joint from a corner should not normally exceed half the general spacing and should preferably be less, due account being taken to ensure stability of the construction. A common cause of cracking in masonry occurs when returns are ignored in determining the centres of movement joints, also the effect of building/wall shape.

Fired-clay masonry

As a general guide unrestrained or lightly restrained unreinforced walls such as parapets and non-loaded spandrels built off flexible membrane-type d.p.c.s will expand 1 mm/m during the life of the building due to thermal and moisture movements. The width of movement joints is governed by the compressibility of joint fillers and the performance of the external sealants. The width of the joint in millimetres should generally be about 30 per cent more than the distance between movement joints in metres, but if in doubt the sealant/joint filler manufacturers should be consulted. Thus movement joints at 12-m centres will need to be approximately 16 mm wide. The above guidance may be conservative for some bricks such as London Stocks which have a lower movement than 1 mm/m.

Expansion of normal storey height walls is generally less than 1 mm/m and Foster and Johnson (1982) have suggested that three categories of movement might be considered dependent on the brick manufacturers recommendations, the higher value perhaps being more appropriate for dark coloured bricks. The values suggested were as follows:

High global expansions   0.9–1.3 mm/m
Medium global expansions   0.6–0.9 mm/m
Low global expansions   <   0.6 mm/m

In general expansion reduces with increasing restraint. However, in unreinforced walls spacing between movement joints should never exceed 15 m in order to avoid cracking due to thermal contraction. Closer spacing may be necessary for walls such as parapets, etc. Where bed joint reinforcement is used spacing of movement joints in excess of 15 m may be satisfactory.

Vertical movement of unrestrained walls would appear to be of the same order of magnitude as horizontal movement. Clay masonry should never be rigidly bonded to concrete or calcium silicate masonry.

Calcium silicate masonry

Calcium silicate masonry should, whenever possible, be constructed as a series of panels separated by movement joints at intervals between 7.5 m and 9 m. However, the shape of walls constructed in this type of masonry can be critical and in general the ratio of length to height of panels should not exceed 3:1.

The movement joints (sometimes described as control joints) are basically to accommodate shrinkage and as such do not necessarily need to have a filler or sealant. However, the author would generally recommend that all joints be sealed and filled with appropriate materials to avoid accidental filling which could be unsightly and perhaps cause problems. Movement joints should not normally exceed 10 mm.

In external walls and some internal walls containing openings, movement joints may be provided at more frequent intervals or the masonry above and below the opening may need to be reinforced to restrain movement. Low horizontal panels under windows, etc. may need special attention. Calcium silicate masonry should never be rigidly bonded to clay masonry.

Concrete masonry

Concrete masonry should, whenever possible, be constructed as a series of panels separated by movement joints at intervals of approximately 6 m. Since there are wide variations in the physical properties of different concrete masonry units, some variation in joint spacing may be acceptable. However, the shape of walls constructed in these types of masonry can be critical and in general the ratio of length to height of panels should not exceed 2:1. This type of masonry should be separated from dissimilar materials such as *in situ* concrete and clay masonry.

The comments regarding sealing of movement joints and joint width made under the heading of calcium silicate masonry are equally applicable to concrete masonry although it may, for aesthetic reasons be considered appropriate, in some instances, to use a movement joint wider than 10 mm.

In external walls and some internal walls containing openings, movement joints may need to be provided at more frequent intervals or the masonry above and below the opening may need to be reinforced to restrain movement. Low horizontal panels under windows, etc. may need special attention.

Offsets and junctions

Joints or slip planes at offsets and junctions are recommended as high concentrations of stress build up at these positions due to movement. Short returns should also be avoided as it has been found that where the length of the return is not more than 686 mm cracking is likely to occur. It is suggested therefore that to avoid cracking of this kind the length of the return should be not less than 686 mm (Figure 4.11).

Parapets

Joints in parapet walls should follow the lines in the main structure and should be carried through the parapet. If additional joints are considered necessary these should be positioned approximately midway between those running throughout the full height of the building. Stability must be carefully checked in such situations, particularly as no tension is permitted at the d.p.c. level.

It is important not to bridge movement joints at the top of free-standing walls with heavy coping stones which may be dislodged.

Single-storey framed buildings

Thermal movement of single-storey buildings and the top storeys of multi-storey buildings can be rather critical due to wide variations of diurnal temperatures. This, coupled with the fact that the floors are subject to only minor variations in temperature, can result in a racking action taking place. It is undesirable to rigidly tie walls in such situations and dowel or flat anchor ties are preferred (Figures 4.15 and 4.16). This gives adequate support to resist lateral wind and other pressures and at the same time some degree of flexibility normal to the lateral forces. It is also important to make allowance for movement at eaves level if steel portal frames are used, as sway may induce cracking in the masonry.

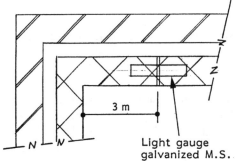

3 m

Light gauge
galvanized M.S.

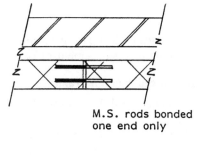

M.S. rods bonded
one end only

Figure 4.15 Movement joint with lateral restrain

Figure 4.16 Movement joint with lateral restraint; alternative detail

High stresses tend to build up at door and window openings in long walls, especially in light-framed factory buildings. It is therefore recommended that movement joints or reinforcement of the masonry should be considered in such situations.

4.8 Masonry cladding to reinforced concrete and timber frames

Problems due to differential movement between masonry cladding (usually clay brickwork) and reinforced concrete-framed structures, as well as timber-framed structures, can occur if provision is not made for such movements. The solution to the problem is usually to provide movement joints and/or flexible or special type wall ties capable of accommodating the differential movement of the materials, thus

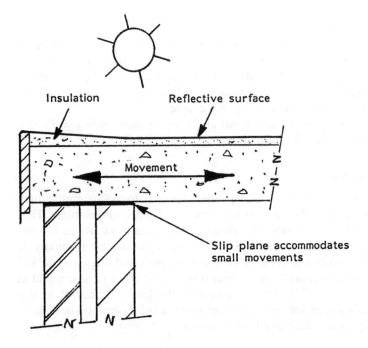

Figure 4.17   Roof movement

avoiding a build-up of unacceptably high concentrations of stress and the subsequent cracking or displacement of components.

Reinforced concrete frames

It was not until the reinforced concrete frame began to supersede the steel frame for multi-storey construction that severe cracking of masonry and widespread shearing of fixings occurred. The reason for this was that steel columns are relatively accurate when erected, do not shrink and have negligible creep, whereas the reinforced concrete frame, even if accurately constructed and adequately cast, does have measurable drying shrinkage, elastic deflection and long-term shortening due to creep. According to Edwards (1966) the drying shrinkage, of concrete can be greater than 0.085 per cent when it is made of highly shrinkable aggregates and when made with aggregates from a wide geological spectrum the percentage can range from 0.066 to 0.085 per cent. In addition to shrinkage concrete columns are subject to creep or plastic shortening, this type of deformation being extremely complex. The change in dimension which occurs due to creep is dependent upon numerous factors, i.e. the magnitude of the applied load, the duration of loading, the age of the concrete, the humidity and moisture content of the concrete, the type of aggregate used, the type of cement used and the amount of water used during the mixing of the concrete. According to Foster (1971), Hollington has assessed creep of concrete subjected to a stress of 10.3 N/mm^2 as 300 microstrain or 0.03 per cent but Foster also points out that this figure could be doubled if a highly shrinkable aggregate is used. Elastic strain of the column is dependent upon the intensity of loading, the quality of the concrete and the amount of reinforcing steel included and can be calculated using simple formulae.

The deflection of beams should not be ignored when infill panels of brickwork are used in framed construction and in addition to elastic and plastic deformation of the beam, where appropriate, shuttering for *in situ* flooring should be struck prior to infilling with brickwork or blockwork, thus allowing the full dead weight to be imposed on the beam and hence the appropriate deflection to occur. The author has seen newly erected walls collapse before any superimposed load has been applied when this precaution has not been taken.

Design
It is essential to make allowance for any differential movement which may occur between the various materials of construction if damage to the structure is to be avoided. To recapitulate, allowance must be made for the following movements in a vertical direction:

(a)   Thermal expansion and contraction of the masonry
(b)   Permanent and cyclic moisture movement of the masonry
(c)   Accidental transfer of load from one leaf of masonry to the other by rigid wall ties (Figure 4.10)
(d)   Drying shrinkage of the reinforced concrete frame
(e)   Elastic and plastic shortening of the reinforced concrete frame
(f)   Deflection and creep of reinforced concrete beams.

In the horizontal direction it is also essential to consider thermal and moisture movement in the cladding. For large structures movement should also be considered in floors, roofs and indeed in the reinforced concrete framework.

To prevent clamping of masonry panels between their supports horizontal movement joints should be provided at the top of each panel. The thickness of the actual joint is usually arranged to equal that of the mortar joint, i.e. 10 mm approximately but the required thickness should be calculated to ensure that this allowance is adequate.

It is essential that the provision for movement should be effective and to ensure that the 'movement joint' is not accidentally filled with debris, a suitable material should be selected as a backing filler for the mastic facing joint. The backing material should be resilient and easily compressible to allow movement to take place. Certain types of fibre board or similar materials may be suitable for construction joints (where shrinkage is the criterion) but they are not suitable for expansion joints. Tests have shown that in some instances a stress of over 2 N/mm^2 can be reached to achieve 50 per cent compression of the material and on removal of the load, the material does not return to its previous dimension. It is not sufficient to merely insert two courses of a proprietary bituminous felt d.p.c. in the hope that this will accommodate the movement (Figure 4.18).

Support for the masonry, particularly when in the form of simple nibs or extensions to the main beams should be accurately detailed to coincide with the masonry coursing. Figures 4.19 and 4.20 illustrate how brickwork has on some occasions been

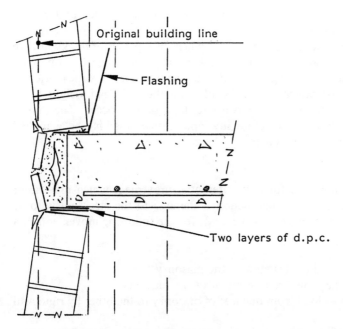

Figure 4.18   Spalling of brick slips

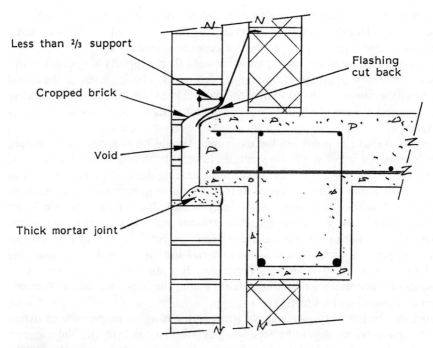

Figure 4.19   Misaligned nib support

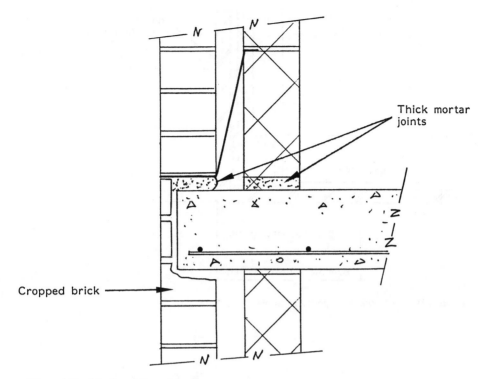

Figure 4.20   Misaligned floor slab

cut in an alarming manner when brick coursing and the nib support did not coincide. It could be argued that this is a problem of poor workmanship but the designer should always locate the supports accurately to ensure that adjustments are unnecessary on site. Whatever the reasons for the cropping illustrated, this is an extremely dangerous practice and must not be tolerated.

Figure 4.21 shows a corner detail where the supporting nibs stopped short and the brickwork returning round the corner was unsupported for the whole height of the building while the remainder of the brickwork was tightly clamped. The author has seen examples of this detail which needless to say resulted in both horizontal and vertical cracking of the brickwork and also cracking of some supporting beams near ground level.

When walls overhang their support it is important that adequate bearing remains and the rule suggested by the old LCC regulations is still considered good practice, i.e. the wall should not be allowed to overhang its support by more than one-third of its actual thickness. If floor slabs or supporting nibs are disguised to give the impression of a building having a continuous sheath of brickwork, it is vital that the designers ensure that the brick slips or sections used to disguise the horizontal number are securely fixed. Figures 4.22 and 4.23 illustrate mechanical methods of support proposed by Foster (1971). Brick slips have been successfully cast *in situ* with floor slabs and beams but considerable care is needed to avoid unsightly stain-

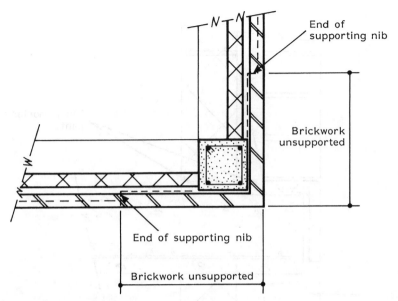

Figure 4.21 Brickwork unsupported at corner

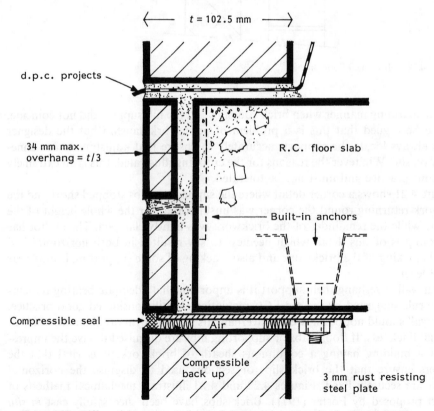

Figure 4.22 Differential movement absorbed. Suggested construction by Foster (1971)

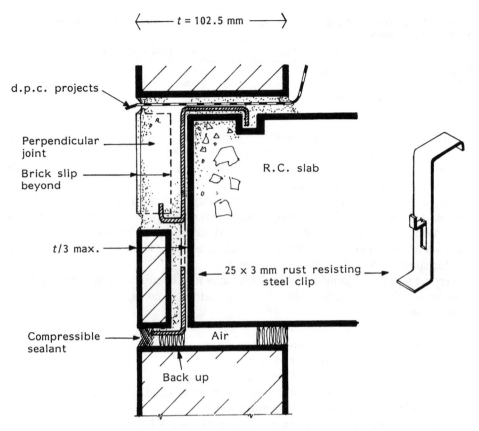

$\longleftarrow$ $t = 102.5$ mm $\longrightarrow$

d.p.c. projects

Perpendicular joint

Brick slip beyond

R.C. slab

$t/3$ max.

25 x 3 mm rust resisting steel clip

Compressible sealant

Air

Back up

Figure 4.23   Differential movement absorbed; alternative detail suggested by Foster (1971)

ing of the surface. Alternatively, a reliable adhesive such as an epoxy resin based mortar might be used. Anything less substantial should be avoided.

Timber frames

When masonry is used to clad timber-framed construction it is assumed that the timber frame supports all the load (both wind loading and dead/superimposed vertical loading) and that the masonry cladding merely transmits the wind loading via the flexible wall ties to the frame. The actual amount of lateral load carried by the masonry depends largely on its geometry. Nevertheless it is important that the masonry is not accidentally loaded due to differential movements of the masonry and the timber frame which might result in high concentrations of stress and subsequent problems.

The largest movements take place due to the drying of the timber, the effect of compressive loading on the frame joints and long-term deformation of the timber (creep).

Shrinkage occurs almost entirely in the horizontal components of the frame and the total movement could be as much as 10 mm at the eaves level of a two-storey house and 15 mm in a three-storey house. Clay brickwork will tend to expand and

TABLE 4.1 **Recommended design movement gaps to accommodate relative masonry and timber moisture and thermal movements**

	Clay masonry	Concrete and calcium silicate masonry
Up to first floor	10 mm	5 mm
Up to second floor	15 mm	5 mm
Up to third floor	20 mm	10 mm

concrete and calcium silicate masonry to contract. It is therefore necessary to make provision for movement at each location where an element supported by the timber framework bridges the masonry, e.g. at eaves, verges and cills. The recommended allowance for movement is indicated in Table 4.1.

It is necessary to separate lintels for the masonry from the timber frame. Similarly, door and window frames should not be fixed to the masonry although this is unlikely to be a critical factor at ground floor level. Movement gaps between the top of the masonry and the underside of the roof structure need not be filled provided the joint is adequately protected by the eaves overhang.

Vertical movement joints and slip planes may also be necessary in long lengths of masonry on the same basis as described for traditional construction.

In summary,

(1) Allow for (a) thermal movement
                (b) moisture movement
                (c) differential settlement
                (d) differential movement of frames and components.
(2) Consider the provision of movement joints at:
    (a) intersecting walls, piers, floors, etc. (Figure 4.24)
    (b) window and door openings (Figures 4.25–4.28)
    (c) change in height or thickness of the wall (Figure 4.29)
    (d) chases in the wall (Figure 4.24)
    (e) movement joints in the structural framework or in the floor slabs (Figure 4.31).
    Alternatively, consider reinforcement of the masonry (Figures 4.27 and 4.28).
(3) Ensure the movement joint is correctly spaced and positioned; also that it is filled and sealed with the correct materials.
(4) Consider the geometry of the walls and the recommended limiting dimensions, as well as the degree of restraint imposed on the construction (Figures 4.25 and 4.26).

Position of movement joints at special features

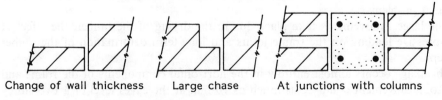

Change of wall thickness   Large chase   At junctions with columns

Figure 4.24   Changes in wall sections

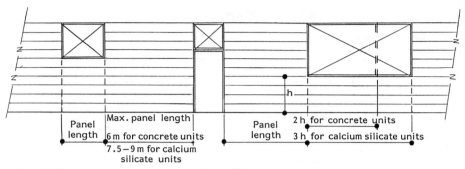

Figure 4.25   Openings in concrete and calcium silicate masonry

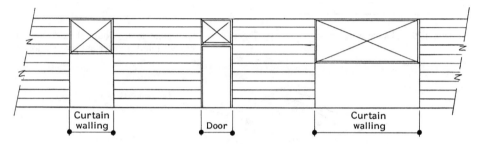

Figure 4.26   Preferred solution for windows in concrete and calcium silicate masonry

Concrete and calcium silicate masonry

The use of reinforcement to control cracking

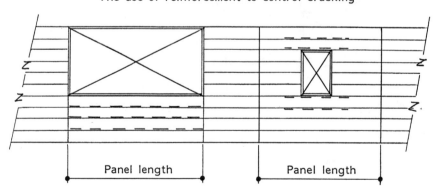

Figure 4.27   Reinforcement distributes
stresses to ends of panel

Figure 4.28   Reinforcement distributes
stresses within panel

(5)   Select the mortar, wall ties and damp proof courses which will give optimum performance and minimize the danger of cracking.

(6)   Do not use short returns or plan forms which will encourage cracking – curved walls, etc. require much more frequent movement joints (Hammett and Morton, 1991) (Figures 4.11 and 4.32).

Movement/control joints

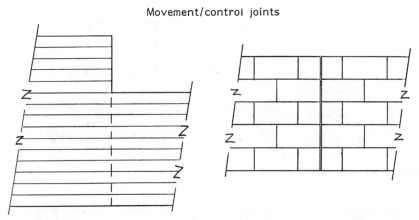

Figure 4.29   Change of height

Figure 4.30   Movement/control joints should always be straight joints. (Never use toothed joints.)

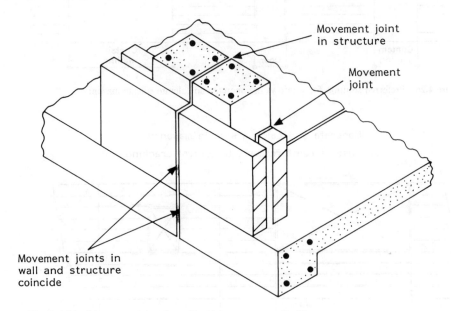

Movement joint
in structure

Movement
joint

Movement joints in
wall and structure
coincide

Figure 4.31   Movement joints in wall and structure to coincide

4.9   Tolerances – support details

Support details for brickwork cladding to multi-storey reinforced concrete and steel structures are frequently suspect, even when the construction is within the accepted tolerances specified in BS 5606: 1990. When bricklayers are presented with large misfits they either take it upon themselves to build walls out-of-plumb and make good at each floor level, or they are instructed to do so by a third party. It is only when the 'faulty details' fail to perform satisfactorily that questions are asked and another building becomes the subject of costly litigation.

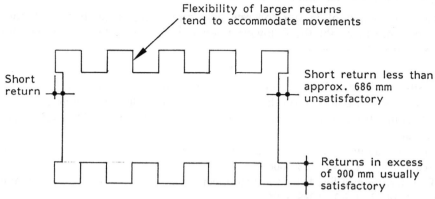

Figure 4.32   Plan of corrugated-type building

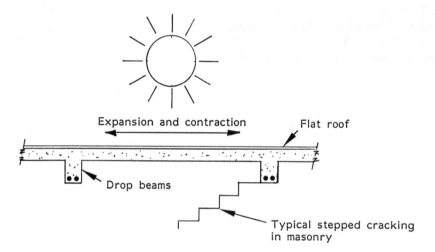

Figure 4.33   Roof movement

Brick slips – BS 5606: 1990, Clause 4.5 states:

'It is common practice in framed buildings for external walls to be built as brick panels supported on the floors. In some cases in this form of construction the elevation will show as a continuous brick face concealing the frame and in such cases the walls should oversail the edges of the floors. The use of brick slips to be fixed to and to hide the edges of the floors is deprecated, as inaccuracies of line at the edges and of the position on plan, of successive floors relative to one another are such that consistent satisfactory bearing for the oversailing brickwork is unlikely to be achieved'.

The following example illustrates the magnitude of tolerances to be accommodated in this form of construction.

Example of calculated tolerances to BS 5606: 1990

For the construction of *in situ* concrete nibs supporting facing brickwork four storeys or more in height the following tolerances should be taken into account when preparing design details:

(a)   Position in plan, calculated tolerance on position of concrete nibs.
Assuming that grid lines on the upper floors are established from grid lines on the ground floor using a theodolite the following deviations must be taken into account:

Transferring grid line from ground to third floor (see Table 3, T.3.3 of BS 5606)	=	±5 mm
Transferring grid line from third to upper floors (see Table 3, T.3.3 of BS 5606)	=	±5 mm
Setting out nearest reference line at each floor level (see Table 3, T.3.3 of BS 5606)	=	±5 mm
Position of face of nib relative to nearest reference line (see Table 1, T.1.5 of BS 5606)	=	±12 mm

Therefore tolerance on position of nib

$$= [(5^2 + 5^2 + 5^2 + 12^2)]^{\frac{1}{2}} \quad = \quad ±14.8 \text{ say } ±15\text{mm (Figure 4.34)}$$

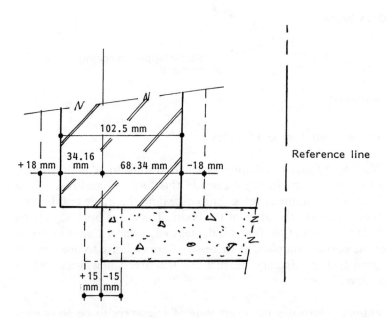

Figure 4.34 Brickwork 2/3 support. (Variations are indicated by the dotted region.)

(b) Position in plan, calculated tolerance on position of facing brickwork.
Position of face of brickwork relative line
(see Table 1,T.1.5 of BS 5606 (Figure 4.35)   =   ±10 mm
Total tolerance on position of nib reference
(see (a) above)   =   ±15 mm
Variability of bearing of brickwork on concrete nib
$= [(10^2 + 15^2)]^{\frac{1}{2}}$ ±18 mm
Similarly it can be shown that for a steel-framed building the
variability of bearing of brickwork on a steel nib or concrete
cased steel beam
$= [(10^2 + 13.5^2)]^2$ =   ±17mm

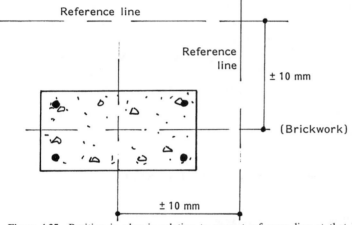

Figure 4.35 Position in plan in relation to nearest reference line at that level – structural frame columns. (Table T.1.5; BS 5606.)

In addition to the above, designers are required to consider the verticality of walls i.e. up to 10 mm for walls of height up to 3 m (Figure 4.36). Also the variation from target plane of concrete beams, i.e. ±23 mm for pre-cast concrete and ±22 mm for *in situ* concrete (Figure 4.37).

Figure 4.38 shows the variability of an *in situ* concrete supporting nib in relation to the brickwork face, and Figure 4.39 the extreme cases based on the above calculations. Both the extreme cases shown in Figure 4.39 illustrate the impractability of brick slips for disguising the face of the concrete nibs. BS 5628: Part 3: 1985, Clause 27.3 states 'Where the use of slips is unavoidable, designers should pay attention to the tolerances on the materials and components to ensure correct alignment of the concrete face or nib, both horizontally and vertically, with the floors above and below. Reference may be made to BS 5606: 1990 which quotes characteristic accuracies for various materials and components'. Clearly if the tolerances calculated above are accepted and brick slips are to continue to be used, a lot more thought needs to be given to methods of fixing the slips, perhaps with flexible brackets capable of taking up the tolerances and the loads which may need to be sustained by such brackets.

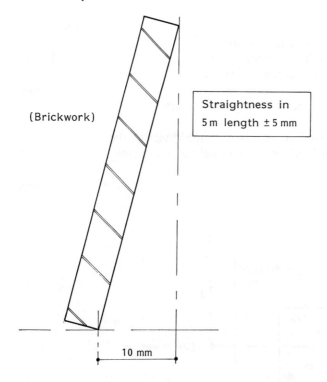

Figure 4.36 Verticality of walls up to 3 m high. (Table 1 T. 1.3; BS 5606.)

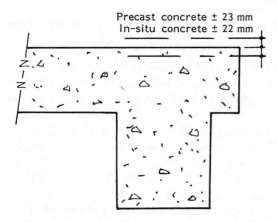

Figure 4.37 Variation from target plane (beams).(Table 1 T. 1.3; BS 5606.)

BS 5628: Part 3: 1985, Clause 27.3 further states 'It is important also that the wall above does not overhang its support by more than one-third its width, e.g. 34 mm for a 100 mm wide leaf. The accurate positioning of the face of concrete in relation to the eventual finished face of the masonry is critical and needs close attention at all stages of design and construction'. Figure 4.39b illustrates that with the extreme calculated

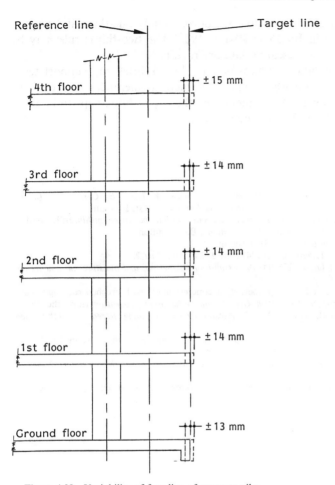

Figure 4.38 Variability of face line of concrete nibs

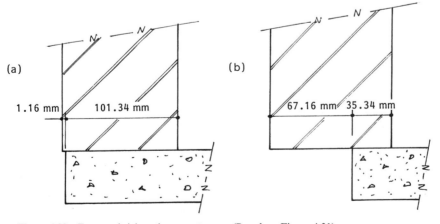

Figure 4.39 Extreme brickwork support cases. (Based on Figure 4.31)

tolerance the brickwork is only supported by approximately one-third and not the required two-thirds quoted in BS 5628: Part 3: 1985. The two-thirds rule may be conservative and perhaps more research data are required.

Based on the above calculations, unless new brickwork/brick slip support techniques can be developed to absorb normal construction tolerances it is difficult to disagree with the BS 5606: 1990 statement, i.e. 'the use of brick slips to be fixed to and to hide the edges of the floors is deprecated'.

References

Beard, R., Dinnie, A., and Richards, R, (1969) Movement in Brickwork. *Trans. Br. Ceram. Soc.* **68**, 73.

BS 5628: Part 1: 1978 *Structural use of masonry – unreinforced masonry.* BSI, London.

BS 5628: Part 3: 1985 *Use of masonry – materials and components, design and workmanship.* BSI, London.

BS 6213: 1982 *Guide to the selection of constructional sealants.* BSI, London.

BS 5606: 1990 *Guide to accuracy in building.* BSI, London.

Building Research Establishment, Digest 165. (1974) *Clay Brickwork, Part 2.*

Building Research Establishment Digest 228. (1979) *Estimation of thermal and moisture movements and stresses: Part 2.*

Edwards, A. G. (1966) Shrinkage and other properties of concrete made with crushed rock aggregates from Scottish sources. *Journal of the British Granite and Whinstone Federation* (now the *British Quarrying and Flag Federation*), Vol. 6, No. 2, Autumn. (Also *Building Research Establishment Digest,* **35**.)

Foster, D. (1971) Some observations on the design of brickwork cladding to multi-storey RC framed structures. *BDA Technical Note,* Vol. 1, No. 4, September.

Foster, D. and Johnson, G. D. (1982) *Proc. Brit. Ceram. Soc.* LBB7-3 p.1–12. *Design for Movement in Clay Brickwork in the UK.*

Hammett, M. and Morton, J. (1991) The design of curved brickwork. *BDA Design Note,* No. 12, November.

5

Bricklaying and blocklaying under winter conditions and frost attack

5.1 Bricklaying and blocklaying under winter conditions

The difficulties of building masonry under winter conditions are well known and there are times when it is prudent to halt the building process and await more favourable conditions. In the UK this is rarely necessary and if certain basic precautions are taken work can usually continue.

Bad weather creates management problems but with efficient site pre-planning and a realistic understanding of the problem, winter working is a challenge to the enthusiastic builder.

The problems of winter building are basically two-fold, (i) working under wet conditions; and (ii) working under freezing conditions. Perhaps the most uncomfortable for the building worker is a combination of the two. Wet weather is a hazard which can occur in any season but during the winter months (November to March) up to between 25 per cent and 50 per cent of the total working hours can be affected by wet weather depending upon location. Indeed, in Great Britain throughout the year an average of one working hour in five is affected by rainfall.

In cold weather mortar sets and hardens much more slowly than at normal temperatures. Even when the air temperature drops to 10°C the setting and hardening is noticeably reduced. When the temperature falls still further to freezing point, the setting and hardening processes cease. If newly laid mortar is allowed to freeze before it has time to harden, the expansion of the water as it turns to ice disrupts the mortar joints and even though setting and hardening will start again when the temperature rises and the ice thaws, the jointing materials will be weak and porous and may need to be discarded.

Delays and difficulties due to cold weather can be substantially reduced if the following precautions are taken:

Weather forecasts

Masonry is particularly vulnerable to freezing during the first few days after laying. Even though the brickwork or blockwork may have been laid during mild weather a sudden frost can cause extensive damage.

Early warning of low temperatures is therefore essential if the work is to be protected in time. Forecasts for change from the Weather Services and general weather forecasts for the next 24 hours, with an outlook for a further 24 to 48 hours are given on radio, television and in the press. The forecasts are for large areas; forecasts for local areas are available via British Telecom Weatherline. The Meteorological Office can prepare special forecasts for a particular building site and comprehensive details are readily available from The Director General, Meteorological Office (MET 0 3b), London Road, Bracknell, Berks RG12 2SZ. In addition a climatology service known as CLIMEST is available to the builder and provides information on average variations in weather as distinct from the actual weather at a specified time – this is useful for planning. Also a service known as METBUILD which gives an actual monthly downtime summary, i.e. the number of hours per working day when weather-related operations could not be carried out. For further details contact the above address or telephone Bracknell (01344) 420242, Ext. 2278.

Working temperatures

Temperatures should be checked regularly adjacent to new masonry and the work protected to prevent freezing for at least 3 days after laying. When night temperatures do not fall below −4°C and day-time temperatures rise above freezing point, insulating material such as sacking, fibreglass or straw quilts, may be sufficient without additional heat. The insulating material must of course be covered with polythene or similar impervious sheeting materials to maintain it in a dry condition. For more severe conditions it will be necessary to supply heat in addition to coverings and insulating quilts.

It is important that sturdy and easily read thermometers be used to measure the temperature of the air and materials. In addition to the normal maximum/minimum type of thermometer for measuring air temperatures, sites should also have robust soil-type thermometers so that temperatures can be measured and not guessed. These thermometers should be used for measuring the temperatures of sand stockpiles, mixing water and mixed mortar. A careful record of all such temperature measurements should be kept, thus ensuring a regular site procedure.

Storage of materials

Bricks and blocks should be stacked clear of the ground and completely covered with tarpaulins or polythene sheeting, bearing in mind that the drier the materials, the less susceptible they are to frost attack. The protective sheeting should completely cover the stacks on all sides and should be weighted down to prevent the wind blowing them off (Figure 5.1). Where bricks and blocks are used under conditions of artificial heating they should be stored under such conditions for at least 24 hours before use.

Neither bricks nor blocks should be wetted to reduce their suction rate during frosty weather. If sand must be stored out of doors it should be covered immediately after delivery and stockpiled in storage bins or on a firm dry base laid to falls to provide drainage (Figure 5.2). Frozen sand must not be used in mortar.   WHY

Cement should be stored well above ground level on a timber floor or platform, preferably in a dry building, but if this is impracticable it should be completely

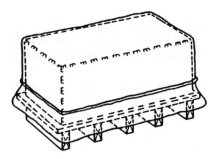

Figure 5.1   Bricks and blocks stacked clear of the ground and covered with waterproof sheeting

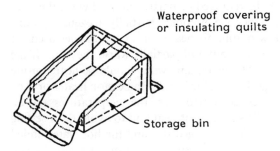

Waterproof covering
or insulating quilts

Storage bin

Figure 5.2   Sand protected from frost

covered with weighted tarpaulins or polythene sheeting. Consignments should be so placed as to permit inspection and used in the order of delivery. Cement affected by dampness should never be used.

Hydrated lime–sand for mortar should be stored on a timber platform or steel sheeting and covered with weighted tarpaulins or polythene sheeting immediately after delivery.

Loose plank covering of materials is not advisable. Careless storage of materials increases the cost of building, because the removal of ice and snow and the thawing of materials used for bricklaying or blocklaying are absolutely necessary before construction may be commenced.

The cost of material protection is small relative to the total contract cost and this fact should not be ignored.

Mortar

Mortar mixes weaker than 1:1:6 cement:lime:sand should not generally be used externally in cold weather and it is recommended that unless a stronger mortar is specified for structural masonry or work below ground level damp proof courses, a 1:6 cement:sand mortar with an air entraining plasticizer be used. (Some clay brick manufacturers recommend stronger mixes for their products. If in doubt consult the manufacturer.) This will have the advantage of earlier setting than a mix containing lime and overcome the problem of maintaining lime-putty in a frost free condition. Mortar plasticizers, which entrain air in the mix, now provide an alternative to or may be used in conjunction with, lime. The air bubbles introduced into the mortar by the plasticizing agent serve to increase the volume of the binder paste, filling the voids in the sand and this correspondingly improves the working qualities with less water to expand on freezing and hence less danger of disrupting the mortar before it has fully cured.

The writer is not aware of any documented evidence that commercially available plasticizers cause efflorescence but some claims have been made to this effect. Certainly, domestic detergents should not be used as mortar plasticizers as many of them contain sodium sulphate which could contribute to efflorescence.

The use of aerated mortars with dry bricks can in some instances, affect bond and some authorities recommend additions of water-retaining additives, such as cellulose ethers to counteract this. Stockpiles of sand should be protected as discussed on p. 112 and this protection should only be removed or partly removed when the sand is being taken from storage. If the stockpile becomes partially frozen, a simple polythene or tarpaulin tent with a low output air heater or a coke brazier used for 24 hours will generally be sufficient to thaw out and partially warm the sand. If the temperature is expected to fall below $-4°C$ the mixing water should be heated using a thermostatically controlled bottled gas burner, an electric immersion heater or even a simple brazier. It is important to exercise control over the temperatures and an appropriate range would be $50-65°C$. To save loss of heat and to avoid flash setting when using hot water the materials should be gauged dry and the hot water added last. Small quantities of sand can also be heated by spreading it on corrugated metal sheets over a heater or by banking it around pipes in which fires have been lit. Having produced a satisfactory mortar for winter working it is important to keep it warm and this can be achieved by gently warming it on a metal sheet over a heater before use, remembering to turn it over once or twice to ensure even warmth. The mortar should then be used as soon as possible if it is not to cool too quickly, especially in windy weather.

To ensure a uniform and intimate mix of the cementing materials and sand, a mechanical mixer should be used where possible. When mixing plasticized cement: sand mortars or masonry cement mortars, care should be taken not to add too much water at the start, as these mortars become more fluid as air is entrained. The old-fashioned rolled-type mortar mill is unsuitable for aerated mortars because it tends to 'roll out' the entrained air. Prolonged mixing of these mortars in other types of mixer can lead to excessive air entrainment and subsequently weak mortars.

In the UK the major suppliers of ready-mixed lime:sand for mortar normally include an air entraining agent in their mixes during cold spells. It is important that users of this excellent quality–controlled product should not gauge admixtures to improve frost resistance as overaeration can only be harmful to the mortar and finished masonry.

Additions of frost inhibitors based on calcium chloride should never be used in mortar for jointing as, apart from being ineffective (i.e. there is no evidence as far as the writer is aware that sufficient heat can be generated in a normal mortar joint to depress the freezing point), they cause deliquescence with the subsequent danger of corrosion of embedded steel and an increased possibility of efflorescence.

Extended curing time

As mortars take longer to gain strength at low temperatures every care should be taken not to load the brickwork or blockwork too soon. When the air temperature is 5°C for example, mortar may take two or three times longer to reach the required strength than it does under normal weather conditions. In addition, an extra day should be added for each day on which the temperature falls below freezing point.

Rendering and plastering

Internal rendering and plastering does not usually suffer damage from frost action, provided that cold winds are kept out and the walls themselves are not extremely cold when the rendering or plastering is applied. Whenever possible all windows should be glazed or covered with polythene and during cold spells warm-air heaters should be placed in the room the day before rendering or plastering so that the walls and materials may be brought to a reasonable temperature. Heating should continue for at least 48 hours after completion of the work. Large temperature differentials should be avoided as this tends to cause crazing in the finished rendering or plaster-work. Temperatures should also be kept above freezing point for lightweight plasters during the application and hardening period. Dry lining will, of course, avoid all the difficulties inherent in cold weather rendering and plastering.

External rendering should not be carried out during frosty weather.

Protecting completed work

Brickwork and blockwork should be covered as the work proceeds to protect it against freezing for from 3 to 7 days, depending upon conditions; cold winds and draughts can be very damaging to new mortar. Special care should be taken with single-leaf walls as they are more readily attacked by frost than thicker walls, especially when exposed on both sides.

One simple method giving nominal protection is to put a close course of bricks on top of the wall at the end of each day, letting them project about 50 mm on either side; then cover at least the work carried out during the past 24 hours with polythene sheeting or similar waterproof covering (Figure 5.3). Additional insulation may be provided under the covering in very cold weather. During exceptionally cold weather it is advisable to use heating to ensure that the masonry is unaffected.

Small protective enclosures with forced air heaters may be suitable and strong, windproof enclosures (Figure 5.4). When bricklaying or blocklaying inside a building all openings should be sealed and where heat is needed portable space heaters are convenient.

Walls should not be heated on one side and remain unprotected on the other. Enclosures should be arranged to allow a circulation of warm air on both sides of the wall.

Protection against rain and snow

Individual materials and recently completed masonry should always be protected against rainwater and snow. The precautions for individual materials are similar to those described above for frosty weather. It is essential that newly completed

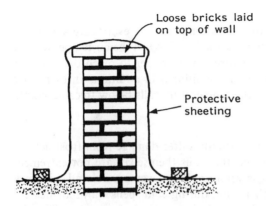

Figure 5.3   Masonry protected after laying

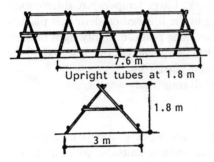

Figure 5.4   Example of scaffold frame for bricklayer's tent

masonry is protected at the end of each day's work against rainwater or in any period of interruption through rain. It is particularly important that perforated or hollow units should be covered during intervals of construction. In multi-storey construction or where scaffolding is used, it is important that the plank adjacent to the masonry and any mortar boards are turned back at the completion of each day's work to avoid splashing of the finished surface.

Rain-soaked masonry takes a long time to dry, especially in the winter months and this often means delay in plastering and decorating. Care should be taken therefore to ensure that any masonry known to be saturated at the time of frost should also be properly protected to prevent damage by freezing.

A simple test to determine if new masonry is frozen is to apply a blow-lamp to the mortar joints. Any apparent softening of mortar is a certain indication that the material is frozen. Under these conditions work must stop until such time as normal drying can continue.

Any work affected by snow or ice due to the coverings being displaced, should be thawed with live steam, a blow-lamp or blow-torch carefully applied (the latter should never be used in conjunction with veneered timber frame construction). The heat should be sustained long enough to thoroughly dry out the masonry. If it is frozen or damaged, defective parts should be replaced before the work continues.

5.2   Individual materials

Bricks and blocks

Bricks and blocks should be stacked and protected as described on p. 112. It is unlikely that many bricks used in the UK will have an excessive suction rate during the winter months. If clay bricks do have a high suction rate and the bricklayer considers they will be difficult to lay, the bricks must not be wetted during cold or frosty weather. To overcome the problem of high suction a 'water-retentive' mortar should be used, i.e. a mortar containing a proportion of lime which will improve its water retentivity and resistance to suction. Calcium silicate bricks and concrete bricks and blocks should never be wetted to reduce their suction rate and it is desirable that a highly water-retentive mortar be used as an alternative to wetting the units.

It is not usually necessary to heat bricks or blocks before laying under winter conditions in the UK but in exceptionally cold weather there may be some advantage in doing so. Recommendations have already been made regarding bricklaying indoors on p. 113.

Fired-clay bricks are sometimes delivered 'hot from the kiln' – such bricks should not be built into the work until at least 7 days have elapsed from the time of drawing from the kiln. The reason for this recommendation is to avoid problems of moisture expansion. This subject is discussed in more detail in Chapters 1 and 4.

Cements

All cements should be stored as described on p. 113. Cement should always be kept dry prior to use but should never be heated before gauging with mixing water.

Masonry cement

Masonry cements marketed in the UK contain approximately 75 per cent ordinary Portland cement and 25 per cent of a fine inert filler such as a ground limestone and an air-entraining agent. British masonry cements tend to contain a higher proportion of ordinary Portland cement than those of many other countries and for this reason it is unnecessary to gauge additions of ordinary Portland cement to obtain mortar of relatively high strength. The strength of masonry cement mortars in the UK is, in fact, only controlled by altering the proportion of sand, the normal range of volume proportions being from 1:3 to 1:7 masonry cement:sand. It is very important that the manufacturer's recommendations on mix proportions should be strictly followed. As masonry cement contains a plasticizer additional air-entraining agents must not be incorporated as over air entrainment could cause serious difficulties. The author has visited numerous sites where troubles have occurred due to misuse of masonry cement. These are usually where the operative on site has assumed masonry cement to be ordinary Portland cement. The results of such misuse can be responsible for frost attack of the mortar; frost/sulphate attack of the mortar and substandard mortar strengths. In each case such malpractices have necessitated the pulling down of the work and rebuilding, a tragedy which should never have occurred had the site staff followed the manufacturers recommendations.

Sand

Poor sand should never be used, particularly in cold weather. Very fine sand has a particularly high surface area and if used in weak mortar mixes or in conjunction with masonry cement it has the effect of diluting the ratio of cement to sand and making the resulting mortar more vulnerable to frost and sulphate attack. Excess pigment can have a similar effect, particularly those based on carbon black. Dirt or loam in the sand will make the mortar weak and may also slow down the rate of hardening of the cement. It is therefore essential that only clean sand is used.

During periods of continuous heavy frost stock piles of sand should be maintained at an approximate temperature of 24°C using one of the following methods:

(a) Passing steam through lances or coils inserted in the sand.
(b) Passing heat from bottled gas burners through pipes laid in ground beneath stock piles or bins.
(c) Erecting a protective enclosure over the sand and heating the enclosed area with suitable space heaters.
(d) Using electric surface heaters or electric blankets (i.e. reinforced PVC insulating quilts which include heating elements)

Small quantities can be heated by spreading the sand on corrugated metal sheets over a heater or by banking it around pipes in which fires have been lit.

Damp proof courses and flashings

Some damp proof course materials (particularly those composed of bituminous materials) tend to harden and become unworkable in cold weather. To overcome such difficulties these materials should, if possible, be stored in a warm dry atmosphere prior to use. Alternatively it may be necessary to heat the materials but a blow-lamp, unless recommended by the manufacturers, should never be used. Where flashings require to be shaped (e.g. to form trays, etc.) great care should be taken during cold weather to ensure that they are not damaged, as slight tears tend to extend with some materials due to creep (not only lead) and extensive damage can occur requiring extremely expensive remedial action later.

Mortar mixes

For suitable mortar mixes reference should be made to Chapter 3.

Summary

(1) Anticipate bad weather by using the meteorological services.
(2) Protect brick and block stacks against wetting.
(3) Use a grade III mortar (1:1:6 cement:lime:sand or 1:6 cement:sand plasticizer) or stronger if necessary.
(4) Heat the sand and mortar in extremely cold weather.
(5) Keep the cement and hydrated lime dry.
(6) Do not use very fine or dirty sand.
(7) Do not use calcium chloride.

(8) Do not wet the bricks or blocks prior to laying.

(9) Never use frozen materials.

(10) Keep the finished masonry above freezing point for at least three days after laying.

5.3 Frost attack of masonry

In the UK, night frosts are common even in mild winters and it is important to protect both the units and the newly constructed masonry adequately both from saturation by rainwater and from frost as already discussed.

It is also important to protect newly finished plastered and rendered finishes during frosty weather.

Exclusion of rainwater

Frost attack of masonry is less likely to occur if rainwater is excluded from the construction and the units are protected prior to erection of the walls. The work may become saturated directly by rainfall or indirectly due to rising damp from below d.p.c. level or laterally from retained materials in retaining walls, etc.

External masonry is much less likely to become saturated if suitable overhangs are provided such as projecting eaves and suitably throated cills and lintels. Flexible d.p.c.s and flashings should never be recessed in mortar as this can permit dampness to bypass an otherwise effective moisture barrier. The projection of such d.p.c.s etc. provides a most effective drip and usually sheds rainwater away from the masonry, protecting it from dampness and frequently unsightly staining. Bell mouths to rendering and similar features at the bottom of tile hanging and other cladding above masonry can also provide protection. At ground level frost attack frequently occurs where poor drainage or unchecked dripping overflow pipes allow the masonry to become saturated.

Frost action

Frost action on masonry units is dependent to a large degree on the pore structure of the unit, but no simple relationship has yet been established between pore structure and frost resistance. Porosity is a measure of the total volume of pore space within a body. High porosity does not necessarily indicate a high permeability as the pores may not be interconnected. Some years ago the BRE identified five types of pore in clay bricks which they classified as: (a) channel; (b) loop; (c) blind alley; (d) pocket; and (e) sealed (Figure 5.5). Loop, blind alley and pocket type pores tend to hold air and consequently never fill completely with water which means that space is available for expansion as the water freezes and forms ice. Therefore bricks in which these types of pore predominate may be more frost-resistant. However, other factors may also affect frost-resistance, i.e. high strength units tend to be frost-resistant but there are exceptions and some low strength units have an excellent record of frost-resistance. Absorption is a totally different property to porosity but bricks having a low absorption tend to be of high compressive strength and good frost resistance. However, some bricks of medium absorption and medium strength may have excellent frost resistance!

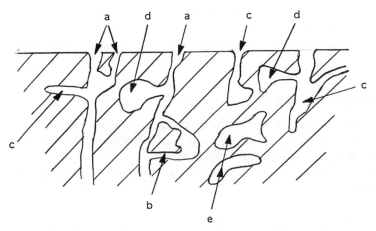

Figure 5.5 Various kinds of pore that may occur in fired-clay bricks. a, channel; b, loop; c, blind alley; d, pocket; e, sealed

Frost resistance

Fired-clay bricks. It is clear from the above that there is no simple test method for assessing the frost resistance of clay products, although freeze/thaw testing of brickwork panels does produce the type of failures found in practice. As there is no test method recommended in any British Standard for assessing frost resistance of fired-clay products the best evidence of ability to withstand frost damage is provided by brickwork which has been in service for some years. BS 3921: 1985 suggests that the manufacturer should provide evidence that bricks of the quality to be offered have been in service under conditions of exposure at least as severe as those proposed for not less than 3 years in the situation in which their use is to be considered and that their performance has been shown by inspection to be satisfactory.

Calcium silicate bricks. Durability and compressive strength tend to be related and experience shows that repeated freezing and thawing has little effect upon the bricks. BS 5628: Part 3: 1985 states that strength class 3 of BS 187: 1978 possess good frost resistance in most applications but higher strength classes are recommended in very exposed situations.

Calcium silicate bricks should never be used in situations where they can absorb salts, i.e. sea-water or adjacent to roads which may be treated with de-icing salts, as deterioration can occur when units saturated in such salt solutions become frozen.

Concrete bricks and blocks. Pre-cast concrete units in general possess good frost resistance, provided they are selected following the recommendations of BS 5628: Part 3: 1985.

Natural stone. Some stones are seriously affected by frost action and care should be taken to ensure that such stones are avoided in positions where they may become saturated. As for other masonry units care should be exercized in the selection of the material and in the detailing of the structure to exclude moisture.

With certain stones, including the softer limestones, the use of thin, open joint construction is not recommended. BS 298: 1972 recommends the use of lead flashings or cornice, string or other projecting courses in order to avoid saturation on horizontal surfaces and subsequent frost damage. BS 5390: 1976 warns that stones suitable for use in inland areas may be totally inappropriate on the coast, being unable to withstand the salt-laden atmosphere and degree of exposure.

Mortars. Chapter 2 discusses the precautions necessary to avoid frost damage of mortar, as well as the inadvisability of using frost inhibitors based on calcium chloride.

Rendering and plastering. Chapter 8 discusses the precautions necessary. Newly plastered walls do need to be protected and external rendering should not be carried out during frosty weather.

Conditions of exposure

BS 5628: Part 3: 1985 defines the various conditions of exposure and these should be considered when selecting materials. In some parts of Scotland for example, driving rain may occur and 30 minutes later the surface of the wall may be frozen. Such conditions are of course extreme and do not apply to the UK generally but explain why renderings or harling is so popular in Scotland. Vulnerable parts of buildings tend to be parapet walls (even where not in an area defined by the Standard (BS 5628: Part 3) as severe exposure) and below d.p.c. level. Such positions tend to be affected by frost action much more than gable walls above d.p.c. level for example. Gable walls are exposed on one vertical face only and are significantly wetted only when there is driving rain. Driving rain is seldom immediately followed by severe frost so that usually some drying takes place, leaving the units less than saturated. In addition, with a heated building the wall surface will be at least a degree or two above the temperature of the cold air and heat stored in the structure will delay the freezing of water in the wall. However, highly insulated buildings may affect the frost resistance of some walls, e.g. frost can sometimes be clearly detected in gable walls above eaves level where the roof space is highly insulated. Also units previously satisfactory in uninsulated walls may not be satisfactory, from a frost resistant point of view, when cavity walls are highly insulated.

Free-standing boundary walls are also more vulnerable to frost attack particularly when brick-on-edge cappings are used without a d.p.c. under the capping units. Designers are recommended to study Table 13 of BS 5628: Part 3: 1985 when considering the form of construction to be adopted from a durability point of view.

If in doubt always seek the manufacturers advice before specifying.

References

BS 5628: Part 3: 1985 *'Use of masonry - materials and components, design and workmanship'*. BSI, London.
BS 3921: 1985 *Clay Bricks*. BSI, London.
BS 187: 1978 *Specification for calcium silicate (sandlime and flintlime) bricks*. BSI, London.
BS 298: 1972 *Natural stone cladding (non-loadbearing)*. BSI, London.
BS 5390: 1976 *Stone masonry*. BSI, London.
Climatological services. Meteorological Office .

Harding, J. R. and Smith, R. A. (1986) 'Bricklaying in winter conditions'. *BDA Building Note 3*, January.
METBUILD – a service to the construction industry. Meteorological Office.
Standard Practice for Winter Working. BEC.
The MET Office – services to the construction industry. Meteorological Office.

6

Salts and stains

6.1 Efflorescence on masonry

Efflorescence consists of deposits of soluble salts formed on the surface of masonry (particularly certain types of clay brickwork). They usually show as loose white powder or as feathery crystals. Occasionally they appear as a hard glossy deposit covering and penetrating the unit faces. Efflorescence can occur on internal as well as external surfaces and it is often a mixture of different deposits.

Efflorescence is most noticeable on new clay brickwork and is generally a temporary springtime occurrence as the new work dries out for the first time. It is sometimes renewed in the second spring of a building's life but is usually much less marked than the first outbreak. Although efflorescence is unsightly it is usually harmless unless the salts crystallize just below the surface, causing disruption of the surface in a similar manner to that due to frost attack. This form of crypto-efflorescence is fortunately much less common than surface efflorescence.

Causes of efflorescence

For efflorescence to occur three conditions must prevail in the following order:

(1)   the masonry must absorb water (Figure 6.1a)
(2)   the masonry must contain soluble salts (Figure 6.1b).
(3)   the water containing the soluble salts in solution must be able to evaporate at the surface of the wall to leave a deposit of salts (Figure 6.1c).

The reason that efflorescence frequently occurs on newly constructed masonry is that water absorbed during the building process tends to upset the balance of salts within the units and these are subsequently transferred to the drying surface. Soluble salts may have come from:

(a)   the masonry (i.e. the units and/or the mortar)
(b)   soil in contact with the masonry
(c)   contamination with sea water or spray

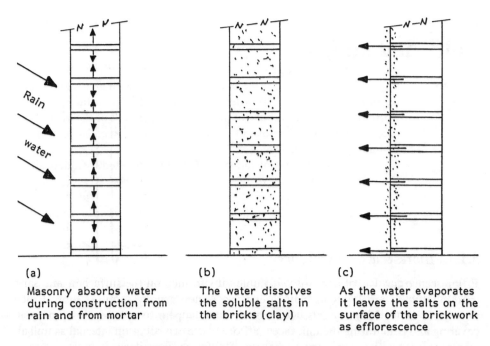

(a)	(b)	(c)
Masonry absorbs water during construction from rain and from mortar	The water dissolves the soluble salts in the bricks (clay)	As the water evaporates it leaves the salts on the surface of the brickwork as efflorescence

Figure 6.1 How efflorescence forms

Very small amounts of soluble salts can cause heavy efflorescence on external masonry but this disappears harmlessly within a few weeks under the combined effects of wind and rain. Efflorescence may damage plaster and paint or emulsion finishes if these are applied before the walls have dried out. This is particularly likely to happen in conjunction with lightweight plasters. The remedy is to make sure that walls are reasonably dry before plastering or that walls are dry lined. On rare occasions when crypto-efflorescence occurs (i.e. the salts crystallize just below the surface) disruption of the surface can occur and if plastered this may be rejected. If the latter occurs a thin layer of the unit is always found attached to the plaster.

Sources of efflorescence salts

Most clay bricks contain water-soluble salts which may contribute to efflorescence, but it can also be caused by salts not originally present in the bricks themselves. The mortar is also a potential source of efflorescence salts and the units or the masonry can be contaminated by salts drawn up from the ground; normally the mortar makes only a minor contribution.

Concrete and calcium silicate units as well as natural stone are less prone to efflorescence than clay bricks and when affected the salts are frequently less visible than on the darker backgrounds associated with clay brickwork.

Salts in clay bricks

The main source of soluble salts found in clay bricks are gypsum and pyrites in the clays from which the bricks are made and sulphur when the bricks are fired using coal. BS 3921: 1985 gives the following categories of efflorescence based on a sample test procedure:

Nil — No perceptible deposit of salts

Slight — Up to 10 per cent of the area of the face covered with a deposit of salts but unaccompanied by powdering or flaking of the surface.

Moderate — More than 10 per cent but not more than 50 per cent of the area of the face covered with a deposit of salts but unaccompanied by powdering or flaking of the surface.

Heavy — More than 50 per cent of the area of the face covered by a deposit of salts and/or powdering or flaking of the surface.

BS 3921: 1985 then states: No specimen shall show efflorescence worse than Moderate.

Mortar

The fine aggregate may occasionally contain soluble salts. Unwashed sea sand always contains salts which are liable to absorb moisture in damp weather. If used it must be thoroughly washed before being introduced into mortar. In certain districts ashes and clinker aggregates are successfully used although they contain sulphates. Lime and Portland cement also contribute to efflorescence in mortar, largely in the form of 'lime bloom'. Research has shown that on bricks where efflorescence has been produced this tends to increase as the proportion of cement in the mortar increases.

In a dense mortar the water tends to dry out through the bricks rather than the mortar and consequently there is increased efflorescence on the bricks and none on the mortar.

Contamination of units

The most likely cause of contamination of units after manufacture but before use is by stacking them on ground made up with ashes or other materials containing salts. Contamination can also occur in buildings where chemicals are used and where fertilizers, salts for de-icing roads and similar substances are stored. Second-hand bricks previously plastered or from buildings where there has been rising damp, or from chimneys, will also contain salts.

Preventing efflorescence

It is not always possible to avoid efflorescence completely because it is affected by the type of unit and by slight and sometimes unpredictable variations in conditions but it can usually be minimized if the following points are observed. Efflorescence cannot occur without moisture being present.

Damp proof courses. Rising damp and water percolating into the masonry should be avoided by the correct selection and use of damp proof courses and flashings. Parapet walls and chimney/roof intersections, etc. need special attention.

Facing units. These, particularly some types of clay bricks with a high soluble-salts content should not be used in very exposed positions. If considering an unfamiliar unit seek the manufacturer's assurance that it is not liable to develop efflorescence. Protect units and masonry against contamination by salt-bearing materials during the construction process.

Mortars. These must have the necessary properties for the work under construction. Chapter 3 deals with mortar mixes for different types of construction and exposure and Chapter 4 gives advice about winter brick and block laying. Air-entraining plasticizers are sometimes added to mortars; quite small amounts are normally used and they are unlikely to contribute to efflorescence. Domestic detergents for dishwashing should never be used as a substitute for plasticizers.

Storage. Keep the units dry. The units should be stacked clear of the ground and completely covered with tarpaulins or polythene sheeting. Dunking of certain types of clay bricks may be necessary if they have a high suction rate but such bricks should never be saturated. It is usually preferable to select a water retentive mortar. Concrete and calcium silicate units should never be wetted prior to laying.

Protection of work. Completed work should be covered promptly so that it does not get thoroughly soaked during construction. One method is to put a close course of units on top of the wall at the end of each day, letting them project 50 mm or more either side. The work is then covered with polythene sheeting or other waterproof covering.

Removal of efflorescence

Surface efflorescence is usually washed away by the rain and no treatment is required. To accelerate the process the masonry can be dry-brushed repeatedly until the soluble salts stop crystallizing. If a wire brush is used care should be taken not to damage the natural texture of the masonry. Efflorescence occurring in a sheltered place where it is not washed away by rain can be removed by repeated washings with a hose, allowing intervals for the salts to be deposited and brushed off between washes.

Persistent efflorescence usually indicates abnormal water penetration of the masonry. Correct any faults in construction; leaking rainwater downpipes and faulty damp proof courses or flashings are common sources of trouble.

Where odd units have disintegrated they should be cut out and renewed with matching units having a low soluble sulphate content. When general disintegration of the wall face has occurred the only satisfactory treatment is to render the surface or in extreme cases, to rebuild the wall.

Efflorescence on internal plasterwork should be lightly brushed off before applying paint or other finishes. On rare occasions when there is general spalling of the internal plaster the source of dampness must be remedied and all the plaster over the affected area removed and the wall left to dry out before it is repaired. Replastering directly on to the affected masonry is not recommended; something is required to break the capillary contact between masonry and plaster. One method is to nail corrugated pitch or bitumen lathing to the inside. This forms a capillary break and a mechanical key for the plaster. Alternatively battens and a dry lining can be used.

Chemical treatment

There are no chemical treatments that can be recommended to neutralize or destroy efflorescence. Surface treatments aimed at suppressing evaporation are likely to be harmful. They can cause decay by forcing the salts to crystallize below the surface with consequent spalling of the unit face.

Efflorescence in old buildings

Efflorescence is usually associated with new brickwork as previously discussed, why then does efflorescence occur on some old buildings?

There can be numerous reasons why efflorescence persists and perhaps the most common is walls built without damp proof courses. The masonry often has pronounced efflorescence in the lower part of the walls and the salts may have been driven from the soil. It is possible however, for such efflorescence to consist of salts originally present in bricks and mortar, the absence of a damp proof course merely allowing the continuous passage of water, drawn up by capillary action which enables the salts to be brought to the surface.

Another common cause of efflorescence is retained earth – either deliberately or accidentally retained. The latter is frequently stacked against the wall and bridges the d.p.c. – this is sometimes noted internally under suspended timber floors where site concrete has not been provided.

In mortar the fine aggregate may occasionally contain soluble salts. Unwashed marine sand always contains salts which are liable to absorb moisture in damp weather. In certain districts ashes and clinker aggregates are used and these may or may not be suitable as they contain sulphates.

Lime and Portland cement also contribute to efflorescence in mortar, largely in the form of 'lime bloom'. One building known to the author has a black mortar which turns white every spring – this is almost certainly due to 'lime bloom' which has reoccurred for the last 30 years. Research has shown that on bricks where efflorescence has been produced this tends to increase as the proportion of cement in the mortar increases.

In dense mortar the water tends to dry out through the bricks rather than the mortar and consequently there is increased efflorescence on the bricks and none on the mortar.

Cleaning of old brickwork using water jets saturates the walls and can, in some instances, upset the balance of the salts in the bricks and cause efflorescence for 2 or 3 years in a similar manner to new construction. One old building known to the author did not have problems with efflorescence until after it had been cleaned!

Leaking gutters and rainwater downpipes are a frequent cause of efflorescence in old buildings also damaged flashings. Parapet walls and free-standing boundary walls are frequently built with cappings rather than copings. This form of construction can be satisfactory with some units but not with others where efflorescence and/or staining of the walls becomes an ongoing problem.

Efflorescence on internal surfaces is a frequent problem for the reasons outlined earlier, and when solid external walls permit moisture penetration. However, damage to decoration is often due to impervious decorative finishes being applied in an attempt to solve the problem of salts forming on inner surfaces with the net result that the paint surface is rejected. If the problem is to be resolved obviously the source of moisture must be eliminated and until the wall has completely dried out (this might take some time) the only satisfactory decorative finish will be one which provides a fine porous surface which can be cleaned if the efflorescence continues. Alternatively some form of dry-lining which includes a cavity may be preferred. If plaster is rejected due to efflorescence this uncommon type of failure is likely to be due to the presence of magnesium sulphate crystals forming behind a lime plaster. The magnesium sulphate reacts with the lime to give a chemical precipitate in the pores of the brick just behind the plaster. This precipitate acts as a semi-permeable membrane, i.e. it allows water to pass through it and evaporate but it holds back any dissolved salt. A thin layer of brick is always found attached to the plaster.

Surface efflorescence is usually washed away by the rain and no treatment is required. To accelerate the process the masonry can be dry-brushed repeatedly until the soluble salts stop crystallizing. If a wire brush is used care should be taken not to damage the natural texture of the masonry. Efflorescence occurring in a sheltered place where it is not washed away by rain can be removed by repeated washings with a hose, allowing intervals for the salts to be deposited and brushed off between washes. Persistent efflorescence usually indicates abnormal water penetration of the masonry and this must be rectified prior to any other remedial work being undertaken. Where odd units have disintegrated they should be cut out and renewed with matching units having a low soluble sulphate content.

When general disintegration of the wall face has occurred, the only satisfactory treatment is to render the surface or, in extreme cases, to re-build the wall.

Efflorescence on internal plasterwork should be lightly brushed off before applying paint or other finishes. On rare occasions when there is general spalling of the internal plaster the source of dampness must be remedied and all the plaster over the affected area removed and the wall left to dry out before it is repaired.

Re-plastering directly on to the affected masonry is not recommended; something is required to break the capillary contact between masonry and plaster. One method is to nail corrugated pitch or bitumen lathing to the inside. This forms a capillary break and a mechanical key for the plaster.

6.2 Staining of masonry

Staining of masonry occurs for a variety of reasons such as metallic corrosion; mould growth; lime leached from associated concrete and chemical reactions within the masonry. Also disfigurement of masonry can occur either accidentally due to paint splashes, etc. or deliberately due to the act of vandals. Regardless of cause, it is usually desirable to remove stains as soon as possible and we will now consider some of the causes of staining of masonry and suggest possible remedial action to remove or make less unsightly such disfigurement.

Cleaning techniques

The methods of removing stains from masonry described in this chapter have generally been found to be satisfactory. However, the manufacturers of the units should always be consulted prior to attempting to remove stains from their units to ensure that they are in agreement with the techniques proposed.

Stains due to efflorescence

Efflorescence and the necessary remedial treatment (if any) has already been discussed. However, rusty stains, unassociated with embedded metals, can occur on mortar joints and these are usually due to iron salts (especially ferrous sulphate) which may form efflorescence when they first come to the surface but which react chemically with lime when they are subsequently washed over the mortar joints by rainwater. This type of problem can occur when certain types of clay bricks are used, e.g. common bricks made from colliery shale. When brickwork built with this type of common brick is plastered there is usually no trouble with staining of plaster if the undercoat contains lime and/or Portland cement as the alkalinity of these materials precipitates the iron and does not allow it to pass through to the surface with any water there may be in the wall as the latter dries out. However, gypsum plasters do not contain alkaline constituents capable of precipitating iron and their pore structure also encourages the passage of salts to the surface. Examples of heavy staining of lightweight gypsum plasters have occurred when they have been applied to walls of colliery shale common bricks that have become wet during construction – some builders add lime to gypsum plaster to overcome the above problem. One form of a yellowish-green stain that sometimes appears on light-coloured bricks and is easily mistaken for vegetation is due to an efflorescence of a coloured salt containing vanadium.

If efflorescence due to vanadium salts is pronounced it may be removed using a tetra-sodium salt of diamenoethane tetra-acetic acid (EDTA) (50 g/litre) but this is a costly process. The wall should not be subsequently washed down. Alternative treatments using oxalic acid solution and washing soda or sodium hypochlorite or household bleach and washing soda are suggested by the BDA (Harding and Smith, 1986).

Vegetation

Algae, lichens and mosses are common on the external surface of walls, particularly in rural areas. Their appearance is sometimes regarded as desirable, particularly on older properties. Algae and lichens are rarely destructive, although mosses do tend to hold water and can encourage frost damage under certain conditions.

Vegetation on masonry may take various forms. It is not likely to be very noticeable unless the walls are wet for long periods; and it may not be objectionable if it is uniformly distributed, but streaks or patches of vegetation may be considered unsightly. To treat such stains, as much as possible vegetation should be scraped or brushed off during hot, dry weather and the surface then washed over with a dilute copper solution. For more detailed information reference should be made to Buildings Research Establishment Digest No. 139 (1972).

Metals

Embedded iron or steel

Stains caused by the rusting of embedded iron or steel in the masonry are part of a more serious problem and can cause serious cracking of the masonry if the metals are not suitably isolated or adequately protected. Where the stains are caused by run-down of the products of corrosion from ferrous metals independent of the masonry the stains may often be removed. Where the staining is confined to the surface of the mortar joints this can sometimes be removed by scraping or rubbing with a round file or carborundum slip. Where scraping is ineffective, chemical treatment using proprietary solutions of hydrochloric acid or alternative acid based preparations are suggested by the BDA (Harding and Smith, 1986).

Iron in mortar sands

Rusty stains may also be caused by mortar sands that contain particles of ironstone. If stains are seen to flow from particular grains in the surface of the mortar this source of staining is indicated. Re-pointing is the only satisfactory remedy for this problem.

Manganese staining

This type of staining of clay bricks is similar to iron staining but is generally dark brown or black in colour and the treatment is essentially similar.

Copper and bronze stains

Green stains due to the corrosion of metallic copper or bronze such as may often be seen on the plinths of statues, calls for treatment at the design stage as remedial measures are usually relatively ineffective. Rainwater draining from the metal should be channelled away inconspicuously in a manner which avoids spillage over the masonry.

Stains from concrete

White stains, distinguishable from efflorescence by the fact that they do not disappear when the masonry is washed by rain, frequently occurs beneath pre-cast concrete copings, lintels, cills, etc., beneath brick cills laid in mortar and on the face of brick parapets or retaining walls with concrete backings. This staining is caused by soluble free lime in the cement which percolates out of the concrete or mortar in the absorbed rainwater on to the face of the masonry. The resulting deposition is then converted into an insoluble white calcium carbonate (similar to limestone) by the action of atmospheric carbon dioxide on the lime leached from the concrete or mortar. To remove such deposits the masonry must first be wetted. It may then be brushed over with diluted hydrochloric acid to dissolve the deposit and the work finally well washed down with clean water. The concentrated acid should be diluted with about five parts of water. If a permanent solution to the problem is to be achieved the rainwater must be completely excluded from the construction with suitable d.p.c.s and/or membranes.

Cement staining from mortar and concrete

Large projections should be removed with wooden or similar implements to avoid damage of the masonry face. The residue should then be removed using a hydrochloric acid solution as described for 'stains from concrete'. It may not be possible to remove stain caused by pigmented mortars using hydrochloric acid and specialist advice should be sought from the suppliers.

Some hard-fired purple stock bricks are known to produce rusty stains on the mortar joints due to the presence of iron salts in the bricks. It is recommended that the stain be allowed to develop harmlessly in the mortar bed joints and subsequently point up the joints after the work is complete. The staining is not likely to recur.

White stains on and around mortar joints
These stains can be caused by the lime leaching out of fresh mortar; due either to the use of saturated units or to fresh masonry at the top of the wall being left exposed to rain. These stains can be removed by diluted hydrochloric acid but it is better and cheaper to keep the units dry and to cover up new work.

Dirty masonry

Deposits of dirt, grime, soot and smoke are generally the result of long-term atmospheric pollution. Also black encrustations under features such as window cills, etc. can occur due to the lack of run-down of rainwater in such areas relative to the remainder of the wall surface. If these stains cannot be removed by scrubbing with liquid detergent solutions, specialist cleaning contractors will be required. However, it must be borne in mind that if specialist contractors use water-spraying techniques with some type of units this may upset the balance of the salts within the units and result in problems of efflorescence.

Paint and graffiti
Removal of hardened paint, etc. is often extremely difficult as the offending materials usually penetrate the surface of the porous masonry materials. Water-rinseable paint removers to BS 3761: 1970 may be useful but generally it is necessary to seek specialist assistance.

Oil, grease and tar
The BDA (Harding and Smith, 1986) recommend that the heavier deposits should be removed by scraping with wooden or similar implements followed by scrubbing with water containing a suitable emulsifying and degreasing agent. For deeper seated stains, a poulticing technique based on white spirit or trichloroethane is recommended.

Brush-applied treatments using organic solvents may spread the stain and make it worse.

Warning

When cleaning masonry with chemicals it is necessary to follow the guide lines recommended by the manufacturers and/or CIRIA (1981), and for clay and calcium silicate brickwork, the BDA recommendations (Harding and Smith, 1986). Units relying upon pigments for their colour may become bleached if treated with acid solutions and the manufacturers of these products should be consulted on methods of cleaning and/or removal of stains.

Prevention

The majority of staining on masonry can be prevented if the walls are designed and constructed in accordance with the recommendations of BS 5628: Part 3: 1985. Faulty design and detailing and bad workmanship are usually the cause of most of the problems.

6.3 Cleaning of masonry

The purpose of this section is to discuss current methods of cleaning masonry and some of the problems encountered in practice. Cleaning of masonry is generally a job for specialist contractors and if inexpertly carried out can lead to serious problems.

Buildings of special interest

When buildings are listed or are in a conservation area the local authority should be consulted before any work is contemplated. In some conservation areas the local authority may discourage or even prohibit the cleaning of specific buildings on the grounds that such work may change the uniform appearance of the area. Where work on features of historic or artistic importance is being considered, guidance from specialists should be sought.

Regardless of the type of building to be cleaned it is always recommended that a small inconspicuous area of the building be cleaned first using the appropriate technique to establish if the end result is likely to be satisfactory.

Preliminary considerations

The reasons for cleaning a building may be aesthetic and/or for maintenance. In practice, cleaning of a building for aesthetic reasons may reveal the need for subsequent maintenance. When the reasons are aesthetic the intentions are usually to reveal the nature, colour or details of the building; to unify the appearance of the building which has been altered or is to be extended and to facilitate the choice of suitable materials for a proposed alteration or extension.

When cleaning is for maintenance purposes, it is usually to remove harmful deposits from the fabric or to expose defects in order to establish the extent and nature of repairs needed.

Selection of method

It is not possible to recommend a cleaning method suitable for all types of masonry. Each building should be considered carefully in the light of all available methods and of previous experience with similar buildings and the method to be used agreed before tenders are invited. It is essential to identify all materials before selecting a method. In particular, it is important to know whether a building stone is limestone or sandstone.

Buildings usually consist of a variety of surfaces and dissimilar materials, each requiring to be cleaned using an appropriate method, which might mean that more than one method needs to be used. Where treatment of one material may harm another adjacent material some form of protection will need to be considered, e.g. glass is damaged by solutions containing hydrofluoric acid.

Materials to be cleaned

Brickwork

Newly erected brickwork should be cleaned (if necessary) using a different technique to that for mature or old brickwork. Sand-blasting and other mechanical methods should generally be avoided as these techniques can remove the face of some bricks and drastically reduce the life expectancy of the structure not to mention its aesthetic appeal. Figure 6.2 illustrates brick ruined by sand-blasting by a firm claiming to be specialist in conservation! Where brushing is needed as an aid to dirt removal from soft material, e.g. red rubbers, it should be carried out with care, using only fibre brushes.

Clay brickwork. Care should be taken to ensure that the method adopted does not remove the surface of the bricks as some units are, for example, sand-faced prior to the firing process and such units generally have a different coloured clay body. To have patches of the clay body exposed after cleaning of the work is totally

Figure 6.2   Brick ruined by sand-blasting

unacceptable. Neutral liquid detergents are frequently successful in removing grime having an organic binder. If this is ineffective, other chemical cleaning agents may prove more satisfactory. Always follow the manufacturer's instructions and clean a trial area before treatment of large surfaces. When acid treatment is used it tends to remove fine material (including pigment if present) from the surface of mortar joints and produce the effect of accelerated ageing of the joints. Water spray processes may cause efflorescence (saturating the brickwork can upset the balance of any soluble salts in the bricks) and is not generally recommended. Some clay bricks contain soluble iron, manganese or vanadium salts and staining could result from wet methods of cleaning which saturate the brickwork.

When smooth-faced clay bricks have been splashed with mortar or concrete this can be removed using a rotary brass brush on a power tool and the resultant marks then removed using a neutral liquid detergent.

Glazed brickwork. This may be subject to surface discoloration or soiling under the glaze. Generally, surface dirt can be removed using a neutral liquid detergent. Where dirt has penetrated the glaze, this is usually the result of crazing of the glaze or, less commonly, the result of water penetration from behind the glaze. Soiling behind the glaze cannot be removed.

Calcium silicate brickwork. If chemical agents other than detergents are to be used, specialist advice should be sought. Generally this type of brickwork can be cleaned by water spray, air abrasive, water abrasive or chemical methods. Cleaning can remove some or all of the surface but this may not be of concern due to the uniform nature and through colour of the body.

Concrete brickwork/reconstructed stone masonry. These materials can be cleaned using any of the water cleaning methods. However, the water can produce a patchy effect after cleaning if differential erosion of the surface has taken place. If water cleaning is ineffective, air abrasion or treatment with chemical agents based on hydrochloric acid may be used.

With some types of concrete brick erosion of the surface can expose the aggregate resulting in an unacceptable appearance.

Natural stone

Limestone. Water cleaning is the accepted method of cleaning limestones. Sometimes brown staining occurs on the surface of certain limestones, such as Portland stone. This is caused by water-soluble tars which have penetrated the pores of the stonework. However, the stains usually fade with time, particularly when exposed to strong sunlight. Cleaning with steam has little, if any, advantage over the accepted water treatments and superheated steam can cause damage to the limestone. Air abrasion is sometimes suggested to prevent staining, but in practice the same discoloration is likely to occur when the newly cleaned stonework is exposed to rainwater.

Wet or dry abrasion may be used to clean the harder limestones but should not be used on the softer stones. Alkaline chemical cleaners should only be used in exceptional cases e.g. for the removal of graffiti and for softening up heavy deposits of dirt prior to using other cleaning methods. Alkali-based cleaners can leave deposits of harmful salts.

Flint. Flint stones may be cleaned with water in the same manner as limestone and cleaning agents based on hydrochloric acid may be used for cleaning the face of knapped flints after pointing. Hydrofluoric acid should never be used on knapped flints as this whitens the faces of the flints.

Sandstone. Water spray methods on their own are usually ineffective and some wet methods can activate the iron content in ferruginous sandstones resulting in staining. Sandstone may be cleaned using either abrasive methods or chemical agents based on diluted hydrofluoric acid. Steam cleaning is not usually effective unless used in conjunction with a chemical agent. Extra care is needed when cleaning soft (calcareous) sandstones as the binding matrix which holds the grains together contains calcium carbonate. If high concentrates of hydrofluoric acid are used there is danger of the carbonate near the surface dissolving, resulting in loosening grains of sand which will be washed away during subsequent hosing down.

Slate and polished granite. These materials require special treatment.

Delicate stonework. Cleaning of delicate stonework may require special techniques or carefully controlled water treatments, etc. Poultices are sometimes used and these consist of a support medium, such as clay, designed to hold, solvent, or a cleaning agent on the face of the stone. The poultices soften the dirt which can then be removed easily.

Safety measures

Cleaning of masonry buildings is classified as a building operation and all the appropriate safety precautions must be taken. Advice on any safety aspect can be obtained from the Health and Safety Executive, Baynards House, 1 Chepstow Place, Westbourne Grove, London W2; Telephone 0171-221-0870.

Warning

Cleaning of masonry requires the observance of several ground-rules:

(i)   This type of work should only be carried out by experienced firms who offer a complete service in masonry cleaning and restoration. It is essential that those carrying out the work have the correct equipment and materials and that only fully trained staff are used.
(ii)   The correct method of cleaning must be clearly specified and used.
(iii) Protective clothing must be provided and used.
(iv)   Scaffolding and screening must be provided where appropriate to protect the general public.
(v)   Mistakes can be costly, not only in economic terms but also to our national heritage.

Further and more detailed information on cleaning of masonry can be found elsewhere (Harding and Smith, 1986; Buildings Research Establishment Digest No. 139, 1972; BS 3761: 1970; CIRIA, 1981; BS 5628: Part 3: 1985).

6.4   Avoidance of stains on masonry

Most staining seen on modern or even old masonry can be avoided by intelligent design, careful selection and protection of materials, good workmanship and maintenance when necessary. No matter how good the masonry units or mortar, if the materials are abused problems are likely to occur and staining of the work is one of the most likely end products.

Design

Designers frequently select bricks or blocks for their aesthetic appeal and do not give sufficient thought to the physical characteristics of the units. Similarly the mortar mix, its colour and profile.

Free-standing walls without copings, unusual profiles and sloping elevations may frequently produce interesting and even exciting architecture, but unless the work is very carefully detailed and the correct materials specified major problems can occur even if the workmanship is faultless.

Cappings/copings

Cappings produce a well-defined clean line on many new buildings but also all too frequently cappings on free standing walls and parapets encourage leaching of free lime from mortar joints even if a damp proof course (in the form of a continuous flexible membrane) has been included near the top of the wall. Also weathering of the mortar joints at the top of the wall is frequently pronounced and moss and algae often thrive in such locations. Unprotected walls (walls with cappings) are also more vulnerable to efflorescence than walls with copings and roof overhangs (see Chapter 10). Similarly, when clay brickwork is unprotected there is a greater danger of sulphate attack of the mortar joints. Copings in the form of overhanging pre-cast units or similar continuous metal or plastic protection with a throating on the underside reduce the incidence of water run-down and consequently many of the associated problems including staining. However, due to the restricted run-down of rainwater locally (as with window cills, etc.) dirty patches may result directly under such projections. It is not suggested that cappings should be totally excluded and only copings used to protect free-standing walls and parapets, but designers should be aware that the use of cappings increases the risk of staining and other defects and to minimize these the following precautions should be taken:

(i)   The masonry units should comply with the requirements of Section (I) Table 13: BS 5628: Part 3: 1985 also the mortar designation.
(ii)   A flexible d.p.c. should be included a minimum of two courses down and this should project on either side of the wall to form a drip (see Chapter 10).
(iii)   The use of air-entraining type plasticizers in the mortar for the capping units may encourage leaching of free lime from the joints and is therefore not recom-

mended. Household detergents should never be used as plasticizers as they are known to encourage staining.

(iv) The cappings should be covered during construction to protect the mortar joints from rainwater until they are thoroughly cured. Similarly, in winter the work should be suitably protected from wet and freezing conditions (see Chapter 5).

Concrete and limestone features

Masonry, particularly brickwork, can be badly disfigured by careless detailing, particularly at junctions with concrete and limestone features such as window cills, lintels and string courses, etc. Once again this is due to leaching of free lime and in extreme cases this can cause decay of brickwork. Figure 6.3 illustrates brickwork badly disfigured by free lime from a supporting concrete backing wall which has leached out of mortar joints and specifically out of weep holes. Figure 6.4 shows a similar area after some cleaning with an acid solution; note the deterioration of the lower course (brick slips in this instance).

Classic staining will be noted in many old buildings with limestone cills where stools have not been provided at the ends, resulting in unsightly 'wisker' stains at each end of the windows. Similar staining and frequent disintegration of soft brickwork can be seen under circular windows outlined with limestone features.

Leaching of free lime cannot occur if water is excluded or suitable drainage details are provided to channel it away from the walls.

Figure 6.3 Brickwork badly disfigured by free lime

Figure 6.4   Similar area to Figure 6.3 after cleaning with an acid solution

Retaining walls
It is important to avoid water percolating through masonry regardless of whether the wall is structural or merely a garden wall. Many otherwise decorative garden walls are ruined because the moisture from retained earth either transports salts from the ground or free lime from the mortar joints onto the exposed face of the wall and/or salts from the units resulting in efflorescence frequently accompanied by lichen and moss growths. Structural retaining walls are usually tanked and suitable provision made for drainage at the base of the wall. While ideally small garden retaining walls should also be tanked, even simple lining with heavy-gauge polythene sheeting and drainage via weep holes will generally be sufficient to avoid unsightly problems at nominal extra cost.

Non-vertical features
Sloping walls and projecting or recessed panels of masonry are often featured in the architectural glossies and encourage the uninitiated to copy or experiment, this is healthy providing the units selected are satisfactory, the details have been carefully thought out and correctly executed and weathering and water exclusion have been taken into account. Many lessons can be learned by a critical inspection of masonry buildings featured in the architectural magazines of 7 or 8 years ago; particular points of interest will be weathering and run-down of rainwater from large windows and other impervious surfaces. Pattern staining where brick slips have been used to disguise concrete supporting members are usually evident due to cold bridging in

these areas or brick slips have a slightly different colour due to firing of the thinner material if the slips were not made from whole bricks. It is interesting to note that structurally honest masonry buildings, i.e. buildings which express their supporting frames (if present) or members and consist basically of vertical walls or true arched construction, tend to exhibit fewer problems than masonry enveloping a structural frame.

Protection of materials

Many of the problems of efflorescence and staining occur because the bricks or blocks have not been suitably stored on site before building into the work. Units should be stored on a prepared hardstanding and covered with well-secured polythene sheeting or tarpaulins. Sometimes the units are incorrectly stored in the manufacturer's yard before delivery, particularly during periods of overproduction. Cement and lime should be stored off the ground and under cover. Sand and ready-mixed lime:sand stockpiles should be placed on an impermeable base and be protected from excessive wetting or drying out, preferably by the use of polythene or similar sheeting. Materials stored on virgin sites without suitable ground cover may be contaminated by chemicals previously used for agricultural purposes.

Workmanship

Incorrect batching of materials for mortar can lead to variations in colour and strength which may result in masonry with a patchy appearance. To ensure accurate batching it is recommended that gauge boxes or buckets are used and these should be carefully filled without compaction. Accuracy in proportioning mortar mixes is important and good supervision is necessary to ensure that mortars are not undergauged with cement.

If the units are multi-coloured or come from different batches they should be mixed to avoid a patchy appearance or specific demarcation lines where different deliveries have been built into the work. Newly built masonry should be protected by covering the wall heads at the end of the working day, or during breaks in construction, when rain is likely. Inner scaffold boards should be turned up at night to prevent splashing and covers should be propped away from the walls so that rain is shed clear, but allowing air circulation to occur to assist drying out. When frost is likely, additional sheeting should be used to cover all freshly built work, and dry hessian or similar insulating material placed beneath the waterproof top cover (see Chapter 5). All shuttering for concrete should be carefully sealed to avoid any rundown of the wet concrete on to the masonry.

Maintenance

Maintenance of masonry itself is rarely necessary if correctly designed and constructed. However, faulty rainwater downpipes, leaky gutters and unchecked overflow pipes frequently saturate masonry locally and if not quickly repaired result in unsightly staining, efflorescence and moss/lichen growths.

Stains caused by rusting embedded metal can also be avoided if the metal is regularly painted.

6.5   Sulphate attack on mortars and rendering

Sulphate attack on mortar can occur when the tricalcium aluminate (C_3A) constituent present in all Portland cements is combined with sulphates in solution. The
reaction which occurs between the soluble sulphates (i.e. magnesium, sodium and
potassium) and the C_3A creates calcium-sulpho-aluminate (Ettringite) and is accompanied by significant expansion of the mortar. This results in crumbling and disintegration of the joints and in the more severe cases, substantial growth and often
disruption and cracking of the masonry. BRE Digest 165 (1974) quotes vertical
expansions of up to 0.2 per cent in facing brickwork and as much as 2 per cent
in rendered brickwork. Horizontal expansion is rather less but may be more obvious.
Serious attack is rarely noticeable in less than two years.

For sulphate attack to occur therefore three constituents must be present: (a)
Portland cement; (b) water-soluble sulphates; and (c) water in the construction
(Figure 4.4).

Source of sulphates

The most common source of the sulphates is fired-clay bricks but it must be remembered that not all clay bricks have a high sulphate content which is readily 'water
soluble'. BS 3921 (1985) specifies the permissible soluble salt content for the bricks to
be described as 'low in soluble salts' (see Chapter 1), but does not prescribe an upper
limit for soluble salts in bricks considered to be 'normal'. Other types of masonry
unit do not normally contain high levels of sulphates unless produced from active
colliery or blast furnace waste, etc.

Other common sources of sulphates are soils, either in foundations or in retained
earth particularly where ground water flow occurs. Some industrial atmospheres can
be highly aggressive and can give rise to sulphate attack. Cases of severe sulphate
attack have been observed when gypsum plaster has been introduced into mortar
mixes either in error or possibly in an attempt to accelerate setting times. Also when
second-hand bricks have been used and old gypsum plaster has not been completely
removed.

Diagnosis

Clay facing brickwork
A two-stage diagnosis is normal:

(a)   The mortar deteriorates, commencing with surface effects similar to those of
frost attack. The mortar will usually whiten and horizontal lamination of the joints
will occur approximately in the centre of the bed joints. Eventually the joints
will become completely friable and it is frequently possible to remove the bedding
material with a finger nail. As sulphate attack can sometimes be confused
with frost attack it is desirable to seek confirmatory evidence of sulphate attack
based on chemical analysis of small samples of the mortar (Figure 4.4).
(b)   If sulphate attack of the outer leaf has occurred horizontal cracking of interior
surfaces may also take place. In solid walls such cracking may coincide with header
courses or in cavity construction where wall ties occur. However, the latter should

not be confused with cracking due to the expansion of corroding vertical twist-type wall ties. If in doubt the wall ties should be checked. Due to the expansive forces in the wall cracking is also likely to take place at wall/ceiling junctions, etc.

Rendered clay brickwork
Diagnosis occurs in three stages:

(a)   Cracking of the rendering in a 'map' or 'crocodile' pattern will be noted. This will have started due to drying shrinkage of the rendering but becomes more pronounced and is followed later by the formation of horizontal cracking along the lines of the mortar joints.
(b)   Wide horizontal and vertical cracks then tend to occur and some outward curling of the edges of the rendering may be noted at the cracks.
(c)   Finally the adhesion of the rendering to the brickwork fails and areas of rendering become detached. Often the brickwork thus exposed will show white efflorescence.

Chimneys
In older type properties with unlined and badly insulated chimneys sulphate attack of the brickwork is more common than with more modern construction.

In addition to expansion on the wettest side of the chimney (the sulphates originating from the clay bricks), sulphates may also attack the rendered lining or joints near the top of the chimney where gases from slow-burning appliances condense on the coldest side of the flue.

The net result can be alarming with chimney stacks distorting as illustrated in Figure 6.5.

Prevention of sulphate attack
Sulphate attack cannot occur if any one of the following is excluded:

(1)   Water in the construction (for a considerable period)
(2)   Sulphates (water-soluble if present in materials)
(3)   Portland cement (unless with a low C_3A content).

Therefore, sulphate attack can be prevented if the masonry is kept dry, by ensuring that sulphates are not present (or they are in insignificant amounts) or by using a sulphate-resistant Portland cement.

BS 3921 (1985) now defines 'low soluble salts' for clay bricks and BS 5628 : Part 3: 1985 makes recommendations for suitable combinations of masonry units and mortar which should ensure adequate durability of masonry in the finished construction for various conditions and situations.

It is clear that only internal walls can be kept completely dry but correct detailing of external walls with adequate d.p.c.s; flashings; roof overhangs and protective membranes where necessary (i.e. for retaining walls), etc. can prevent walls becoming saturated over long periods and hence the risk of sulphates from the units (when present) can be kept to a minimum.

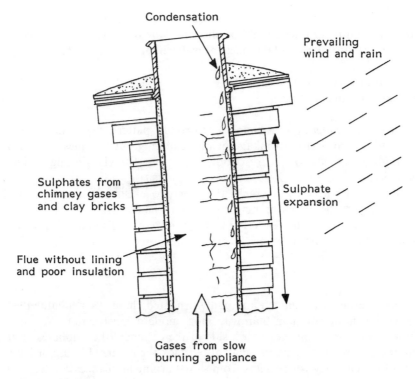

Figure 6.5   Sulphate attack on chimney mortar joints lining render

Parapet and free-standing walls, as well as retaining walls are generally the most vulnerable and designers are recommended to exercise special care and to consult Table 13 of BS 5628 : Part 3: 1985.

Remedial work

Once sulphate attack has been diagnosed it is usually too late to carry out effective remedial action as the structural integrity of the masonry is almost certainly impaired, also the appearance of the building.

If sulphate failure is detected in its early stages and there is unlikely to be any danger due to the reduced structural integrity of the walls the following remedial action could be taken:

(i)   The sources of moisture should be located and measures taken to eliminate further entry into the masonry.

(ii)   The defective masonry should be re-pointed using sulphate resisting cement mortar.

(iii) If rendering has been applied all cracked and non-adherent rendering should be removed, the construction allowed to dry, all loose material and efflorescence removed by dry brushing and the wall re-rendered.

References

BS 3921: 1985 *Clay Bricks*. BSI, London.

BS 3761: 1970 *Non-flammable solvent-based paint remover*. BSI, London.

BS 5628: Part 3: 1985 *Use of masonry - materials and components, design and workmanship*. BSI, London.

BS 6270: Part 1: 1982 *Cleaning and surface repair of buildings. Natural stone, cast stone and clay and calcium silicate brick masonry*. BSI, London.

Buildings Research Establishment Digest No. 139. *Control of lichens, moulds and similar growths*. HMSO, 1972.

Buildings Research Establishment Digest 165 *Clay Brickwork*. 2 May 1974.

Buildings Research Establishment Digest 280 *Cleaning external surfaces of buildings*. December 1983. HMSO.

CIRIA. (1981) *A Guide to the safe use of chemical in construction*.

Code of Practice on Stone cleaning and restoration. August 1974. Federation of Stone Industries.

Harding, J. R. and Smith, R. A. (1986) *Cleaning of brickwork*. BDA Building, Note 2. September.

7
Rain penetration, dampness and remedial measures

7.1 Masonry – exclusion of rainwater

If rain penetration of masonry is to be avoided it is essential to consider carefully the design, detailing, workmanship and materials in relation to the local exposure conditions.

BS 5628: Part 3: 1985 gives recommendations on types of construction and appropriate wall thicknesses for single leaf walls based on exposure categories, but the latter can only be determined when an assessment has been made of the local wind-driven rain index.

Assessment of exposure

The quantity of driving rain impinging on a vertical wall face depends upon the intensity of the rainfall and wind speed. The BRE (1976) have postulated that the quantity of rain falling on a vertical wall surface is proportional to the quantity falling on a horizontal surface and to the local wind speed. In order to determine the wind driven rain index account must be taken of the rainfall in the area under consideration; the wind speed for the area and special features such as the spacing and height of surrounding trees and buildings also whether the ground is flat or steeply rising. Appropriate correction factors are then needed to convert the annual wind-driven rain index to the local annual index.

Based on the above, CP 121: Part 1: 1973 (now withdrawn) defined three categories of exposure as follows:

Severe – severe conditions obtained in area with a driving rain index of 7 or more.

Moderate – moderate conditions obtained in districts where the driving rain index is between 3 and 7 except in areas which have an index of 5 or more and which are within 8 km of the sea or large estuaries in which the exposure should be regarded as severe.

Sheltered – Sheltered conditions obtained in districts where the driving rain index
is 3 or less, excluding areas that lie within 8 km of the sea or large
estuaries where the exposure should be regarded as moderate.

In areas of sheltered or moderate exposure high buildings which stand above their
surroundings, or buildings of any height on hill slopes or hill tops, should be
regarded as having an exposure one grade more severe than indicated by the maps
(Figure 7.1).

Developments since the publication of that code, such as the introduction of
insulation into cavity walls and the advent of improved meteorological data, have
made it necessary to increase the number of exposure categories from three to six.
The basis of calculating the six categories of exposure was originally described in DD

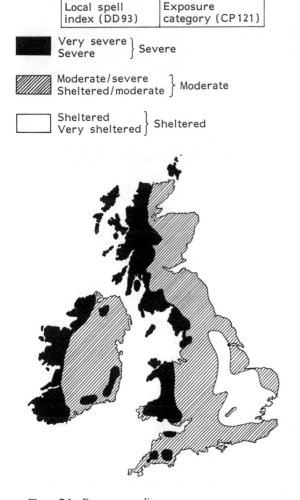

Figure 7.1   Exposure gradings

93 (1984) but this was superseded in 1992 by BS 8104 (1992) and factors to modify rainfall according to local conditions are now based on empirically determined data from experimental sites rather than solely on wind speed data as before. In addition, factors to allow for building shape and size have been introduced also based on data from experimental sites.

BS 5628: Part 3: 1985 defines the six categories in terms of the local spell index as shown in Table 7.1. Where exposure categories overlap the designer is required to decide which is the most appropriate category for his or her building.

TABLE 7.1 Exposure categories

Exposure category	Local spell index (DD 93)	Exposure category (CP 121)
Very severe	L/m² per spell 98 and over	
		Severe
Severe	68 to 123	
Moderate/severe	46 to 85	
		Moderate
Sheltered/moderate	29 to 58	
Sheltered	19 to 37	
		Sheltered
Very sheltered 1 litre/m² = 1 mm of rain	24 or less	

The indices are not precise due to inherently variable meteorological data which are reflected in the above definitions.

Factors affecting rain penetration of masonry

When masonry is subjected to driving rain or panels of masonry are tested in the laboratory using the procedure laid down in BS 4315: Part 2: 1970 the mode of penetration and the degree of water absorption are influenced by the following factors:

(a)   The skill of the bricklayer, blocklayer or stonemason
(b)   The suction rate/absorption rate of the unit
(c)   The permeability of the unit
(d)   The manufacturing process and type of material
(e)   The surface finish of the unit
(f)   Admixtures contained in the unit
(g)   The type of mortar used
(h)   The type of mortar bedding
(j)   Applied finishes to the completed wall, i.e. silicone treatments, renders and special treatments.

Workmanship

The quality of workmanship is extremely important and the skill of the bricklayer/blocklayer is perhaps the most important single factor influencing the permeability of the construction. Voids in joints, particularly perpend joints due to incomplete filling are one of the most frequent sources of leakage reported.

Dense external masonry leaves require more care in the laying process as they tend to be impervious to rainwater which runs down the external face of the wall and penetrates only at the mortar joints. This is known as the 'raincoat effect', in which rainwater tends to penetrate the bed joints by capillary action at the brick/mortar interface unless an exceptionally good bond has been achieved and also via the incompletely filled perpend joints. This type of failure can be accentuated when dense perforated or hollow/cellular units are used.

When fired-clay bricks having a high absorption, e.g. 25 per cent or more are used, the wall acts like a sponge and absorbs rainwater (the 'overcoat effect'). While all mortar joints should be filled, this type of unit tends to be more tolerant of minor imperfections. Rain penetration tests in the laboratory would suggest that this type of wall quickly becomes saturated and is less effective than the denser units. However, generally the reverse is the case in practice, as although the 'sponge effect' readily absorbs rainwater it also quickly releases it when exposed to drying winds and it is only in positions of very severe exposure that the wall is exposed to wind-driven rain for extended periods of time (Figure 7.2).

Mortar

The type of mortar used can have a considerable effect upon the bond and work-ability and it is extremely important that its properties should be compatible with the units used for building the wall. For lower absorption fired-clay bricks the designer should consider using one of the less permeable mortars such as $1:0\frac{1}{4}:3$ or $1:\frac{1}{2}:4\frac{1}{2}$. For other types of masonry unit the selection of mortar is governed by the other factors such as accommodation of movement, durability and strength.

The type of mortar bedding affects the permeability of a wall depending upon the technique of laying, i.e. solid bed, furrowed bed or shell bed (two thin strips of

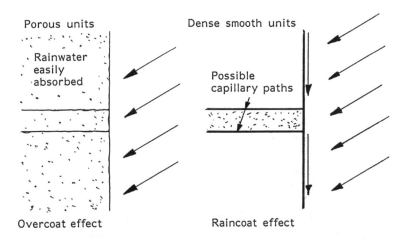

Porous units   Dense smooth units

Rainwater easily absorbed

Possible capillary paths

Overcoat effect   Raincoat effect

Figure 7.2 Effect of porosity

mortar), and there is little doubt that a correctly laid solid bed of mortar produces the best results.

Cement:lime:sand mortars are preferable to straight cement:sand mixes for several reasons, not least of which is their resistance to rain penetration. When mortars containing lime are subjected to cycles of wetting and drying (due to wind-driven rain or sprayed water under pressure in the laboratory) some redistribution of the lime takes place. This phenomenon has been said to be responsible for sealing of hair cracks in mortar joints caused by drying shrinkage of the mortar, hence the term 'autogenous healing'. The writer has tested this theory in the laboratory and obtained confirmatory results.

The use of all admixtures should be strictly in accordance with the manufacturer's recommendations as too little, or perhaps more important, too much of the additive may produce disastrous results. For example, when pigments are used they tend, due to their non-cementitious character and relatively high surface area, to adulterate the mix, reducing bond strength and in some instances compressive strength. Where carbon black is used as a colouring agent, quantities greater than 3 per cent by weight of the cement may affect these properties. On one site visited by the writer a 1:1 cement/pigment had been used and bond between the mortar and brick was non-existent; the resultant capillary path sucked water through the wall with remarkable efficiency when subjected to a simple hose test.

Plasticizers of the air-entraining type improve frost resistance during laying and aid workability but overspecification, particularly with high suction units if not correctly treated, may be the cause of poor bond and consequently water penetration.

Joint finish and profile

The mortar joint finish and profile are important and can influence the degree of resistance to rain penetration. It is essential to fill the joints correctly and tooled or ironed joints such as the concave (bucket handle) or weathered (struck) are more resistant to rain penetration than flush or recessed joints. Recessed joints are not recommended for perforated or hollow units for obvious reasons and the manufacturer should be consulted regarding the suitability of other units which may or may not have adequate frost resistance if used in conjunction with such joints (Figure 7.3).

Wall types

Single-leaf walls

BS 5628: Part 3: 1985 does not recommend (from a rain resistance point of view) the use of even-rendered single-leaf walls in conditions of 'very severe' exposure unless additional cladding is used. Similarly, it does not recommend the use of unrendered masonry in conditions of 'severe' or 'moderate/severe' exposure. Recommended minimum thicknesses for less severe conditions are given for both rendered and unrendered walls.

The resistance to rain penetration of calcium silicate or clay brickwork without rendering or cladding is dependent upon both the thickness and absorptive properties of the units, whereas the rain resistance of dense concrete units is more depen-

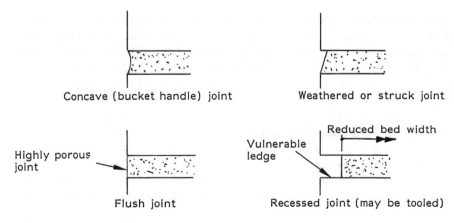

Figure 7.3   Joint profiles

dent on thickness. The use of shell bedding for hollow units may reduce rain penetration to the inner surface.

Unfilled cavity walls
Most thin external leaves of masonry permit some rain penetration under extreme conditions and, particularly when dense units are used for the outer leaf, rainwater may even run down the inner face of the outer leaf. This water penetration is not usually detrimental if effective cavity trays are provided at window openings, etc. so that the rainwater can escape via weep holes (these are usually open perpend joints at 900 mm centres incorporating plastic baffles to prevent wind-driven rain blowing back above the cavity tray on the inner leaf).

In most situations a 50 mm air cavity between the inner and outer leaves is considered satisfactory, but where there is an increased risk of rain penetration wider cavities may be considered. If cavity walls are to be effective the following should be borne in mind:

(i)   Wall ties should be provided at the appropriate centres (usually not less than $2.5/m^2$) and they should slope down slightly towards the outer leaf. They should be corrosion-resistant (generally austenitic stainless steel). It is important to specify the correct type of wall ties to accommodate differential movement of the inner and outer leaves, but at the same time due account must be paid to structural requirements.
(ii)   The cavity should not be bridged by pieces of broken masonry or mortar.
(iii)   Care should be taken to ensure that projections from windows, snapped headers, etc. do not penetrate the cavity.
(iv)   The cavity should never be bridged by horizontal d.p.c.s except under copings.

Filled cavity walls
Filling the complete cavity with thermal insulation material to a large extent defeats the object of cavity wall construction unless the insulation material effectively per-

mits the downward passage of any rainwater without any transfer of moisture to the inner leaf. Even if this can be achieved effectively, wet insulation material must lose a high proportion of its insulation properties. Also some insulants tend to have an objectionable smell when damp. According to BRE Digest 236 (1980) filling the complete cavity with thermal insulation does increase the risk of rain penetration through the wall.

There is little doubt that the risk of rain penetration in cavity wall construction will be reduced if only partial filling is adopted and a wider cavity specified to ensure that a minimum air space of 50 mm is maintained between the inner face of the outer leaf and the insulation material. The use of widths less than this increases the risk of rain penetration and BS 5628: Part 3: 1985 suggests that there should be various restrictions related to the local exposure conditions and the type of construction.

Rendered cavity walls
Rendering of cavity walls considerably enhances their resistance to rain penetration; hence its popularity in Scotland and the West country where the rain appears to fall horizontally at times!

The use of rendered construction does mean that some of the dangers of filled cavities are reduced as far as rain penetration is concerned but the writer would recommend partial filling for new construction.

Movement joints
The resistance of a wall to rain penetration can be impaired due to cracking of the masonry if insufficient provision for movement is made in the form of movement joints.

When movement joints are provided care should be taken to seal the joints with a suitable mastic to prevent rain penetration at the joints.

7.2 Dampness in masonry

In addition to direct rain penetration due to driving rain on the wall face, dampness in buildings frequently occurs due to poor detailing, the use of inappropriate components and bad workmanship.

Horizontal damp proof courses are used just above ground level to prevent rising dampness from the ground and at other positions such as under copings to prevent rainwater percolating downwards. Vertical damp proof courses perform a similar function and should be provided at jambs of openings and in positions where the inner leaf of a cavity wall would otherwise make contact with the outer leaf.

Failure to provide an effective damp proof course, cavity tray or flashing usually results in dampness within the building and the replacement of faulty d.p.c.s etc. can be an expensive operation.

Horizontal d.p.c.s

At the base of the wall in modern construction a flexible d.p.c. is normally selected (these are discussed in Chapter 3). Alternatively, for clay brickwork a minimum of two courses of d.p.c. bricks may be used if laid in a 1:3 Portland cement:sand mortar (again see Chapter 3 for more detailed information). D.p.c. bricks have excellent load bearing properties but should not be used in conjunction with other incompatible materials, i.e. concrete and calcium silicate units; they are not suitable in other positions to prevent rainwater percolating downwards. In many locations it is necessary to form junctions (i.e. at corners) and it is usually preferable to use pre-formed d.p.c. cloaks in these positions.

In cavity wall design the selection and detailing of d.p.c.s should be based on the assumption that rainwater will penetrate the outer leaf of the wall and run down the inside of the outer leaf. Also at the base of the wall it should be assumed that dampness will rise from the ground in both leaves of the external wall (Figures 7.4–7.6). Based on the above the following rules apply generally:

(i)   The d.p.c. must be not less than the thickness of the leaf of masonry it is intended to protect. Flexible d.p.c.s should ideally project approximately 12 mm externally to form a drip which will shed rainwater away from the wall face.

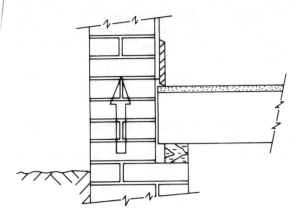

Figure 7.4   Solid wall – no d.p.c. – rising damp

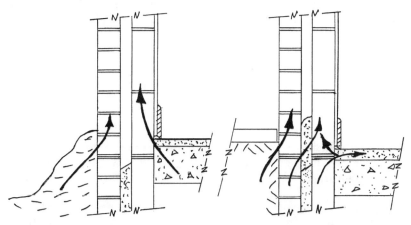

Figure 7.5   Earth and floor bridging d.p.c   **Figure 7.6**   Footpath above d.p.c. Mortar droppings above d.p.c. Porous screed bridging d.p.c

(ii)   Under no circumstances should the d.p.c. be bridged by recessing; pointing over; rendering; plastering or tiling, etc.

(iii)   Joints in flexible d.p.c.s should be lapped at least 100 mm or welted depending upon the material and position. All lapped joints should be sealed.

(iv)   All d.p.c.s should be laid within a bed joint on a smooth bed of fresh mortar unless they are required to accommodate differential sliding movements between the units above and below them. In such cases the mortar bed should be trowelled smooth and allowed to set and then cleaned off before the d.p.c. is laid. It is essential not to use coarse aggregates which might damage the d.p.c. Under no circumstances should d.p.c.s and cavity trays be pierced for services, reinforcement, fixings, etc.

(v)   Weep holes should be provided to drain any rainwater collected in the cavity. They should be positioned immediately above the cavity tray and may be formed by leaving open perpend joints, usually at 900 mm intervals to coincide with coursing of the masonry. Not less than two weep holes should be provided over openings. When cavity filling is anticipated it may be desirable to provide weep holes at closer centres. In areas of severe exposure rainwater draining from the cavity may be forced back and wind uplift, particularly in tall buildings, may force the water up the cavity tray and on to the inner leaf. In such situations it is often advisable to produce a deeper than normal cavity tray, i.e. more than 225 mm and/or to use 'T'-shaped outlet baffle pipes rather than unprotected open perpend joints.

(vi)   Cavity trays should be carefully formed to discharge water to the outer face and, wherever possible, the part of the tray which bridges the cavity should be continuously supported.

(vii)   It is generally necessary to provide cavity trays over all openings and these should project at least 100 mm beyond the end of a lintel.

(viii)   D.p.c.s should be provided under cills unless the latter is completely impermeable. Similarly, they should be provided under copings and the mortar bedding supported where it bridges the cavity.

(ix)   It is important to make provision for drainage of cavity walls below the d.p.c. in the external leaf.

(x)   Some flexible d.p.c.s require very careful handling during cold weather and should be unrolled carefully to avoid cracking. Cracking may also be caused by bending these materials over sharp edges.

Vertical d.p.c.s

Vertical d.p.c.s are required at openings and in all positions where the inner and outer leaves would otherwise make contact.

Flashings

Flashings should be sufficiently malleable to permit dressing into shape but be sufficiently stiff to maintain their shape and to resist lifting by wind forces. They should also be resistant to corrosion.

Cappings and copings

Parapet walls, free-standing walls, retaining walls and chimney terminals should be provided with copings.

All copings whether pre-cast or built up using creasing tiles should have an over-hang to shed rainwater away from the wall and suitably throated on the underside if this is horizontal.

When cappings are used special care is needed in the choice of materials, both for the capping and for the walling beneath. Regardless of whether a coping or a cap-ping is used a continuous d.p.c. is necessary under joints and this should be bedded in a suitable mortar (see Chapter 2). Where cappings are used the d.p.c. may need to be positioned several courses down in order to obtain sufficient weight to prevent unacceptable movement of the capping.

Special care will need to be taken in positions where movement joints occur and these must not be bridged by copings or cappings.

Framed construction

When masonry is supported by a structural frame special attention should be paid to detailing of d.p.c.s and cavity trays to ensure their continuity.

Where the masonry is supported on an edge beam, or on a floor slab, a cavity tray should be used to prevent moisture penetration into the building. If a column, or other special feature, obstructs the cavity wall the cavity tray should be continuous around the member. When a support bridges the cavity a vertical d.p.c. should be included between support and the external leaf and stop ends formed in the cavity tray.

Chimneys and other features

Chimneys and other special features need careful detailing to exclude rainwater. It is essential not only to provide d.p.c.s, cavity trays and flashings but also to select the correct masonry unit and mortar combination. In some positions, such as the junc-tion of a chimney and a pitched roof, and in other exposed positions, it may be desirable to use a different unit from that selected for the remainder of the construc-tion.

Pitched roof/flat roof

Pitched roofs generally give greater protection than flat roofs concealed by parapet walls. This is particularly so if the eaves of the pitched roof has a good overhang. Parapet walls are particularly vulnerable from a durability point of view and in the writer's opinion, should be avoided whenever possible.

Applied external finishes

Greater resistance to rain penetration can be achieved by cladding masonry with metal, plastics, shingles, slates, tiling or timber.

Rendering substantially enhances the rain resistance of masonry if specified and the recommendation of BS 5262 (1976) should be carefully followed.

The use of masonry paint systems to BS 6150 (1982) and other proprietary exter-nal finishes including colourless treatments, e.g. silicone-based water repellents to BS 6477 (1984) may increase the resistance to rain penetration. However, these surface treatments may modify the surface characteristics of the wall, i.e. the surface may shed the rainwater quickly due to the 'raincoat effect' as opposed to the 'overcoat is has been known to cause problems previously not encountered! ments also have a limited life.

ıg damp

Beıv.   treatment for rising dampness is put in hand it is important to confirm that the problem is in fact, rising damp and not due to other causes, i.e. water introduced during construction (new buildings); condensation; plumbing defects; from cleaning and spillage or rain penetration.

The effects of rising damp, as seen in some old buildings where a damp proof course has not been provided or is defective, are fairly classic. The usual signs are: (a) a fairly regular line of efflorescence up to 1.5 m above ground level; (b) discoloration below this line in the wall; (c) a general darkening and patchiness of the surface; and (d) frequently mould growth with its attendant musty smell, loose wallpaper if present or spalling of the painted/emulsioned surface.

Salts brought to the surface up from the ground, i.e. chlorides and nitrates are deliquescent and frequently concentrate in local areas in sufficient quantity to cause dampness due to extraction of moisture from the atmosphere. In modern buildings rising dampness is usually due to the damp proof course having been bridged by the floor screed or rendering, by a pathway or soil stacked against the external wall, or in cavity walls by mortar droppings.

Rising damp frequently occurs in old houses without damp proof courses. It was not until the Public Health Act of 1875 that d.p.c.s became mandatory. Since then d.p.c.s in many old houses have given trouble, either because they were incorrectly installed or have subsequently become ineffective with age.

BS 743 (1970) specifies materials acceptable for damp proof courses and more information on this topic is included in Chapter 3.

Remedial measures

When rising damp is due to bridging of an effective damp proof course by retained earth, paths above the d.p.c., or rendering and pointing, the remedy is usually fairly simple. Mortar droppings at the base of the cavity can generally be removed by cutting out individual units at intervals. However, where no damp proof course exists or where it is defective more radical methods are necessary, these include:

(a) Inserting a damp proof course
(b) Drainage and drying
(c) Chemical/patent systems
(d) Covering or lining.

Before deciding to insert a damp proof course several factors need to be considered, i.e.

(i)   Is the wall stable and without serious defect?
(ii) Is the wall likely to move if opened up?

(iii)   Is the wall of an economic thickness to treat, above 450 mm may be suspect?
(iv)   Are there any obstructions such as service pipes or cables within the wall?

If any of the above are in doubt, professional advice should be sought.

Insertion of a damp proof course
This is undoubtedly the most certain method of curing the problem but may not be the cheapest. Nevertheless, in the long term it frequently turns out to be the most 'cost effective'.

The traditional method of inserting a damp proof course is to cut out one or two courses of brick, a short length at a time and replace with two course of slates laid to break joint, each slate being bedded in 1:3 Portland cement:sand mortar or two courses of d.p.c. bricks, again laid in break joint and in 1:3 mortar (For a full specification see BS 743 (1985) or Chapter 3.)

Some years ago tools were developed for cutting out mortar joints, before the insertion of a new damp proof material. Some of these tools and their method of use are described in detail in BRE Digest 27 (Second Series) (1962). The tools included a chain saw developed at the Building Research Station for this type of work.

Insertion of the various d.p.c.s each have their problems and BRE Digest 27 (1962) warns that small settlements can be expected after insertion of the new membranes.

The damp proof materials inserted in existing walls can be fortified or even replaced with modified mortars incorporating styrene butadiene resin (SBR) or epoxy resins. However, before proceeding with these alternatives or composites the materials should be checked for compatibility and effectiveness with the manufacturers.

Drainage and drying
Where it is impractical to insert a damp proof course, e.g. natural stone walls having rubble cores, measures should be taken to reduce the ground water in contact with the wall or to accelerate evaporation of dampness in the wall. These measures may be effective on their own or may be sensible preliminaries before lining the wall.

Improved site drainage may be the simplest method of lowering ground water by inserting land drains parallel to the base of the wall laid to falls and making good with a porous fill. Alternatively, exposure of the wall below ground level (if this is practical) to encourage evaporation there and/or placing corrugated sheets of non-corrodable material up against the wall to encourage air circulation.

In the past, systems using porous fired-clay pipes have been installed in damp walls, up to two-thirds the thickness of the wall and sloping down towards the external face to drain or siphon moisture. These tubes are often described as capillary tubes and have been shown to be of limited value as they can become hygroscopic due to the deposition of soluble salts extracted from the ground water.

Chemical systems
These are said to provide a water-repellent zone across the wall at the level where they are injected into the masonry. The chemicals used for this purpose include

aluminium stearates, silicone solution in an organic solvent and water-soluble silicone formulations (siliconates) which become water-repellent after curing. The
injected chemicals provide resistance by lining the pores of the masonry and reducing
capillary action up the wall. One method relies on a latex system to seal pores in the
units and mortar matrix.

The aqueous siliconate solutions are drip fed into the wall via perforated tubes in
drilled holes at 150 mm centres which, in turn, are connected to reservoir bottles. The
drilled holes are often made in mortar joints to facilitate simple making good afterwards.

Another system uses frozen pellets of siliconate solution which are popped into the
drilled holes; on melting, the chemical solution disperses and cures to form a barrier
to capillary action. Silicone solutions in solvent and silconate/latex mixtures are
normally injected under pressure in 100 to 150 mm stages, each stage being allowed
to cure before the next is started.

Electro-osmosis
Electrical damp-inhibiting systems have been available for some considerable time
but opinion as to their efficiency varies. The Building Research Station in 1962
stated 'The Station has not been able to establish the effectiveness of this procedure'
(BRE Digest No. 27; 1962).

Passive systems are said to earth the wall by connecting electrodes in the wall to
electrodes of similar metal in the ground, or to produce an electrical potential by
galvanic action using dissimilar metals for the two sets of electrodes.

Active systems (the only truly electro-osmotic systems) induce an electrical potential from an external source between sets of electrodes.

Wall linings
Lining walls internally does not reduce the dampness in the wall. Indeed, because it
reduces evaporation from the wall surface it may encourage moisture to rise higher.
It is therefore recommended that if dry lining is the method of damp control decided
upon, it should be storey height. If however, for economic reasons, it is decided to
limit the height this should be sufficient to overlap the affected area by at least 450
mm.

Several patent methods of dry lining are available but perhaps the cheapest is
plasterboard on timber battens, which also has the added advantage of increasing
thermal insulation. To prevent rotting, the battens should have been pressure-
impregnated with an appropriate preservative (creosote should not be used) and
to reduce mould growth in the cavity between the back of the plasterboard and
the wall surface, the back of the lining board should be treated with a fungicide.

Alternatively, if foil-backed plasterboard is used this will avoid the need for the
fungicide treatment and will increase the thermal insulation. Rust-resisting fixings
should also be used for securing the battens to the wall and a d.p.c. behind the
battens should give added protection.

7.4   Water repellents for masonry

Masonry provides adequate protection from the weather when correctly designed and constructed. However, when rain penetration problems occur property owners frequently look for a cheap solution to their difficulties without identifying and correcting the root cause of the problem. Sometimes a water repellent is the answer. Water repellents will not cure rain penetration in all circumstances; indeed, they may sometimes be ineffective or even detrimental.

Early water repellents were solutions or emulsions of waxes, oils, resins, fats or metallic soaps such as aluminium stearates. More recent water repellents are based upon silicone resins in the form of solutions of silicone in organic solvents, aqueous solutions of siliconates and silicone emulsions (BRE Digest 125; 1971). Monomeric alkyl alkoxy silanes have also recently been proposed for use as water repellents for masonry.

Formulations

The formulation of water repellents are not generally revealed and specifiers should seek the advice of manufacturers on the appropriate product for any specific substrate. However, BS 6477 (1984) does classify the water repellents for masonry in four groups as follows:

Group 1   Predominantly siliceous
Fired-clay surfaces. Sandstones, e.g. Darley Dale, Wealden, St. Bees, Blaxter and Hollington stone. Mature and other hydraulic cement based materials, e.g. cement and cement:lime renderings.

Group 2   Predominantly calcareous
Natural limestone, e.g. Portland, Bath, Hopton Wood, Clipsham, Weldon, Doulting, Caen. Cast stone made using hammer compacted factory processes.

Group 3   Freshly made cementitious materials and other materials of similar alkaline nature
New or repaired surfaces where brickwork or stonework has been bonded, painted or rendered with cement-based materials.

Group 4   Calcium silicate
i.e. Sandlime or flintlime brickwork.

Some repellents are suitable for use on more than one substrate. When treating new work Group 3 repellents should be used unless the mortar is allowed to cure for at least 1 month before application of the material.

Effect of treatment

Water-repellent liquids improve the resistance to rain penetration of the units and the mortar (but not necessarily the interface between the units and the mortar) without appreciably changing their overall appearance. The treatment lines the pores with a water-repellent material which inhibits capillary absorption, so that water tends to remain on the surface (in a similar manner to raindrops on a window pane) until it subsequently runs off. However, although the treatment may be effective during showers, appreciable quantities of rainwater may be absorbed by some surfaces during prolonged rainfall. Because of the repellent nature of the treatment and the subsequent increased run-down of rainwater on less porous surfaces, greater rain penetration can occur due to deficiencies in bond between mortar and units, also at cracks and incompletely filled perpend joints.

Water-repellent treatments do need to be renewed from time to time and BS 3826 (1969) gives information on durability testing. Following treatment the surface remains permeable to water vapour and absorbed moisture can still escape by evaporation but at a slower rate than from an untreated surface.

Efflorescence

Walls subject to efflorescence should not be treated with water-repellents. When a wall contains soluble salts and becomes wet, the salts move in solution to the surface of the wall, the water evaporates from the face of the wall and leaves the salts deposited on the surface in the form of an unsightly but usually harmless efflorescence. Treatment with a water repellent may be detrimental because the salts are prevented from moving in solution to the surface and are retained within the pores at a depth dependent upon the penetration of the water repellent. Consequently, spalling of the treated surface may then occur due to pressure from the build-up of salts.

Surface efflorescence is usually washed away by the rain and no further treatment is required. However, if it is considered necessary to accelerate the process the masonry can be dry-brushed and washed repeatedly until the soluble salts stop crystallizing. It is always advisable to wait until efflorescence has ceased before applying any water-repellent treatment.

Treatment

Preparation

Before application of a water-repellent, cracks wider than hair-cracks should be made good, as should defective mortar joints. Dirty surfaces should be cleaned after consultation with the manufacturer; caution may be needed as some cleaning agents such as detergents may nullify the water-repellent properties of the treatment. Lichens or algae growths should be removed by brushing following treatment with a suitable fungicide.

Selection and application

Care should be taken in the selection of water repellents, as not all are universally applicable to all types of masonry surface and reference should be made to the four groups outlined above. Application should be in accordance with the manufacturer's instructions and any warning about the toxicity or inflammability of the solvent should be strictly observed.

Water repellents should be applied by brush flooding or low-pressure spraying. The masonry face should be traversed by slow horizontal passes, allowing a run down of not more than 150 mm, which will be covered again by the next pass. Generally work should commence at the top of the wall and proceed to the bottom in a continuous operation. The water repellent should not be worked into the masonry by surface agitation.

Water repellents should not be allowed to come into contact with bituminous materials as they tend to dissolve in organic solvents and cause staining of adjacent materials. Similarly, water repellents should not be allowed to come into contact with glazing since they are difficult to remove. All glazed areas should therefore be protected before application of the treatment.

Preservation of masonry

Water repellents should not generally be used as preservatives for decaying masonry. However, an exception to the rule may be appropriate for some stone walls exposed to salt spray, where the external face is sound but the internal surface is blistering, flaking or powdering. However, this is usually more appropriate to stone mullions, jambs and transoms. Here treatment of the external face with a water repellent may significantly reduce the decay internally. On no account should the treatment be applied to the internal surfaces.

Surfaces treated with water repellents (new or newly cleaned masonry) tend to retain a clean appearance longer than untreated surfaces but there is a risk of a streaky appearance if external features channel rainwater down the walls in an uneven manner.

The use of water repellents on the upper part of a wall only may result in an increased flow of rainwater locally which in turn can result in water penetration problems at a lower level.

References

BS 5628: Part 3: 1985 *Use of masonry – materials and components, design and workmanship*. BSI, London.
BS 8104: 1992 *Methods for assessing exposure of walls to wind-driven rain*. BSI, London.
BS 4315: Part 2: 1970 *Method of test for resistance to air and water penetration – permeable walling construction (water penetration)*. BSI, London.
BS 5262: 1976 *Code of Practice for external rendered finishes*. BSI, London.
BS 6150: 1982 *Code of Practice for painting of buildings*. BSI, London.
BS 6477: 1984 *Water repellents for masonry surfaces*. BSI, London.
BS 743: 1970 *Materials for damp proof courses*. BSI, London.
BS 6477: 1984 *Water repellents for masonry surfaces*. BSI, London.
BS 3826: 1969 *Silicone-based water repellents for masonry* (to be withdrawn shortly). BSI, London.
Buildings Research Establishment Digest No. 27 (Second Series) (1962) *Rising Damp in Walls*. October.
Buildings Research Establishment Digest 125 *Colourless treatments for masonry*. January 1971. HMSO.
Buildings Research Establishment Digest 236: 1980 *Cavity Insulation*.
Buildings Research Establishment Report (1976) *Driving Rain Index*.
CP 121: Part 1: 1973 *Walling – Brick and block masonry* (now withdrawn). BSI.
DD 93: 1984 *Methods for assessing exposure to wind-driven rain* (now withdrawn). BSI.

Wall finishes

8.1 Internal plastering

Types of plaster

Lime plaster. This is rarely used nowadays because of its low strength and susceptibility to damage. However, lime gauged with plaster increases its resistance to impact damage and gives a smoother finish, reducing drying shrinkage and increasing the rate of hardening. Each undercoat must be thoroughly dry before application of the subsequent coat. For initial decoration only permeable decorative finishes that are resistant to alkali should be used.

Cement:lime:sand plasters. These are suitable as undercoats on most masonry backgrounds. They give a relatively strong hard surface when used as final coats but tend to shrink on drying if insufficient key is provided or a poor sand is used. This type of finish is suitable for damp conditions and textured finishes (see p. 169). For smooth finishes a wood, felt or cork-faced float should be used as trowelling with a steel float brings excessive laitence to the surface which can result in crazing. As for lime plaster an alkali-resistant permeable decorative finish should be used initially. According to BRE Digest 49 (1964) aerated cement:lime:sand plasters are useful barriers to efflorescence.

Ordinary Portland cement:sand (aerated) and masonry cement:sand plasters. These are similar in hardness to cement:lime:sand mixes having the same cement content.

Masonry cement is defined by the International Organisation for Standardisation as 'a finely ground mixture of Portland cement or other appropriate cement and of materials which may or may not have hydraulic or pozzolanic properties and physical properties such as slow hardening, high workability and high water retentivity which makes it especially suitable for masonry work'.

The mix proportions for masonry cement:sand plasters differ from those of ordinary Portland cement:sand (aerated) plasters. Aerated rendering mixes are particularly sensitive to the grading of the sand and to the suction of the background. The suction of the background removes water from the mix and tends to break down the bubble structure during subsequent working. Sand for aerated mixes should there-fore contain sufficient fine material to help stabilize the bubble structure and main-tain some workability if the air content is reduced. As masonry cement contains more fine material than ordinary Portland cement the requirements for 'fines' in the sand grading may differ dependent upon the type of cement used.

Cement:lime:perlite plasters. These are pre-mixed lightweight aggregate (expanded perlite) with a mixture of ordinary Portland cement and hydrated lime as binder. It is understood that the strength, properties and behaviour of these plasters are similar to 1:1:6 cement:lime:sand plasters but their coverage and thermal insulation proper-ties are better than the traditional sanded plasters. The finishing coat can be formed in this material (using a wood float) or alternatively a Class A or B gypsum plaster.

Gypsum plasters. Class A (Plaster of Paris) is unsuitable for most jobs other than for small repairs as it sets very quickly.

Class B (retarded hemihydrate) plasters give a hard surface and are sufficiently resistant to impact for normal use. They have a controlled set and expand slightly during setting.

Gypsum plaster should not be used on walls which are likely to remain damp after setting of the plaster as this causes weakening and disintegration.

Lightweight gypsum plasters are generally used when it is required to increase fire resistance or to marginally improve thermal insulation or reduce surface con-densation. The plaster is used neat and is premixed, consisting of a lightweight aggregate (e.g. expanded perlite or exfoliated vermiculite) and a Class B plaster. The surface hardness is generally as for Class B plaster but due to the high water content the plaster is slower drying.

Preparation of background materials

Dense clay brickwork. Smooth dense clay brickwork tends to have a low suction rate and provides a poor key for plastered finishes. This type of background tends to expand slightly (see Chapter 1) but has negligible drying shrinkage. To improve the key/bond of the surface the mortar joints should be raked out and the surface to be plastered treated with a bonding agent, spatterdash or an additional key provided in the form of a wire mesh. If a spatterdash coat is used a 1:2–3 Portland cement:coarse sand is suitable and should be allowed to harden before applying the undercoat. Wire mesh should be fixed at least 5 mm clear of the surface.

Pre-mixed lightweight gypsum bonding plaster can be applied without the above treatments.

Common clay brickwork. Medium- to high-suction clay brickwork should have raked out joints unless a purpose-made key is provided in the individual bricks. For high-suction backgrounds it may be necessary to wet the surface with a water brush unless special plasters are used. Movement characteristics are similar to dense clay back-grounds.

Dense concrete brickwork/blockwork and calcium silicate brickwork. These materials are subject to drying shrinkage which varies from low for some calcium silicate units to relatively high for concrete brickwork and blockwork. Most units have a moderate suction rate and provide a reasonable key. However, joints should be raked out unless a key is provided. Some backgrounds of smooth-faced concrete units have low coefficients of expansion compared with that of gypsum plaster and consequently differential thermal/moisture movement may cause problems. Bonding agents of the polyvinyl acetate (pva) will strengthen the bond in these instances.

Concrete and calcium silicate backgrounds. These must be allowed to dry before application of the plaster to minimize the effects of drying shrinkage.

Lightweight aggregate units. These generally have a relatively high drying shrinkage and low suction rate. Because of the exposed aggregate at the surface they usually provide an excellent key for plastering. Walls must be dry to minimize drying shrinkage and some manufacturer's recommend the use of bonding agents or special plasters. Always follow the manufacturer's recommendations.

Aerated concrete blockwork. This has a moderately high drying shrinkage and suction rate. Walls must be dry to minimize shrinkage movement. Some manufacturers recommend the use of a bonding treatment with bonding plaster or browning plaster incorporating a water retentive agent.

Mix proportions

Table 8.1 gives the mix proportions for undercoats; Table 8.2 gives final coat mixes and Table 8.3 suitable mixes for the various backgrounds.

TABLE 8.1 Undercoat mix proportions (by volume)

Reference	Type of plaster	Mix proportions
1	Lightweight browning	
2	Lightweight bonding	Premixed
3	Lightweight cement:lime:perlite	
4	Gypsum (Class B):lime:sand	1:3:9
5	Cement:lime:sand	1:0–$\frac{1}{4}$:3
6	Cement:lime:sand	1:1:5–6
7	Cement:lime:sand	1:2:8–9
8	Portland cement:sand (plasticized)	1:5–6
9	Portland cement:sand (plasticized)	1:7–8
10	Portland cement:sand	1:3

Sand used for gauging with gypsum plasters should be Type 1 to BS 1198. However, if Type 2 sand is used the quantity shown in the Table for gypsum:sand mixes should be reduced by one third.

Defects

Cracking caused by background movement. Cracks frequently occur due to differential thermal/moisture movement of the background materials and the plaster. Cracking usually appears around lintels, window cills and at other junctions of dissimilar materials, e.g. in concrete blockwork walls where clay

TABLE 8.2 Final coat mix proportions (by volume)

Reference	Type of plaster	Gypsum	Lime putty	Sand*
(a)	Lime:gypsum (Class B)	$\frac{1}{2}$–1	1	–
(b)	Gypsum (Class B)	1	$0\frac{1}{4}$	0–1
(c)	Gypsum (Class B)	Neat	–	–
(d)	Lime (neat or sanded)	–	1	0–1
(e)	Lime:gypsum (Class B):sand	$\frac{1}{4}\frac{1}{2}$	1	0–1
(f)	Lightweight gypsum	Neat	–	–
(g)	Portland cement:lime:sand	1	$0\frac{1}{4}$	3
(h)	Portland cement:lime:sand	1	1	5–6
(j)	Portland cement:lime:sand	1	2	8–9

*Sand to Table 2 of BS 1199.

TABLE 8.3 Plaster suitable for various backgrounds

Background material	Plaster final and undercoats*					
	a, b, or c	d and e	f	g	h	j
Common clay brickwork	3 or 8	3, 4, 6, 7, 8 or 9	1 or 3	5	6 or 8	7 or 9
Dense clay or calcium silicate brickwork	3, 6 or 8	4, 6, 7, 8, or 9	1 or 3	5	6 or 8	7 or 9
Dense concrete or closed-surface lightweight concrete blocks	–	–	2 or 3	–	–	–
Open textured lightweight aggregate concrete blockwork	3, 6 or 8	3, 4, 6, 7, 8 or 9	1 or 3	–	6 or 8	7 or 9
Aerated concrete blockwork	6 or 8	4, 6, 7, 8 or 9	2 or 3	–	6 or 8	7 or 9

*See Tables 8.1 and 8.2 for references

bricks have been used as make-up pieces. Whenever possible the use of dissimilar backgrounds should be avoided but if this is not possible junctions should be reinforced with expanded metal. Alternatively, a straight cut should be made through the plaster before it sets along the line of the junction or plaster trims can be used to avoid unsightly cracking.

Shrinkage cracking. This usually takes the form of fine hair cracks on the finishing coat. If gypsum plaster has been used throughout, the cause is either due to the use of a very loamy sand in the undercoat or an excess of lime putty in the finishing coat. If a cement- or lime-based undercoat has been used with a gypsum finishing coat the cause is usually due to the application of the finishing coat before the undercoat has dried and completely shrunk. Alternatively, accelerated drying could have caused the problem by exposure to sunshine, wind or artificial heating and wide temperature variations during the early hardening period. Treatment of these fine cracks is extremely difficult and the most satisfactory solution is usually to cover the surface with wallpaper.

Bond failure. This may be caused by the application of a strong undercoat on a weak background or a strong finishing coat on a weak undercoat. Alternatively, the use of an excessive thickness of plaster; insufficient wetting down of the background or lack of surface preparation. To avoid bond failure: (a) use a plaster mix comparable with or of lesser strength than the background or preceding coat; (b) rake out mortar joints; (c) if available use keyed units or apply a bonding agent or bonding type plaster where necessary and scratch undercoats extensively; (d) allow ample time for drying out before applying plaster and provide sufficient (but not excessive) warmth and ventilation; (e) remove oil, laitance, dust and free water; (f) ensure that the correct suction rate is provided to the background by surface wetting or use of a bonding agent; and (g) if necessary provide a spatterdash or wire mesh. If the latter is used premixed lightweight metal lathing plaster should be used.

Rapid or slow-setting of plaster. Plaster fresh from the factory and still hot can be quick setting (British Gypsum White Book, 1986). The bags should therefore be opened to allow the plaster to cool before use. Conversely, the set may also speed up if the plaster has been kept too long or if bags have become damp. Dirty mixing water or impurities in the mix will also speed up the set.

Blistering. This can occur due to intense local movement of the final coat relative to the undercoat.

Efflorescence. This may emanate from the background (see Chapter 6) of brickwork or blockwork and from certain undercoats based on Portland cement. Dirty sands are also a cause of efflorescence. BS 1191 (1973) limits the percentage of efflorescent salts in gypsum plasters to a negligible amount. The water-soluble salts, when present, can be carried in solution to the surface of the plaster as the construction dries out and appear as efflorescence. The salts become evident on drying out and can be brushed off. Decoration of the walls should be delayed until the drying out process has completed, otherwise it may be badly damaged. If it is not practical to delay decoration until drying out is complete a temporary permeable paint may be used.

Sealed-in water. The presence of sealed-in water from the building process may result in damage to painted surfaces. The recommendations regarding drying out and temporary permeable finishes described above under the heading efflorescence are equally applicable.

Grinning of joints. The marked difference in suction rates of the mortar joints and the masonry units frequently result in the joints being visible (grinning) through the plaster. This can best be avoided by the use of mortar and units of similar character or, in some instances, by the use of an additional coat of plaster. The most satisfactory solution to the problem, when it occurs, is usually to paper the walls.

Rust staining. This is usually due to the use of unsuitable plaster on metal lathing or plaster in contact with corrodible ferrous metal under damp conditions. However, it occasionally occurs as a result of iron stains from mortar joints or manganese staining from certain clay bricks (see Chapter 6). To avoid staining from metal lathing, etc. a metal lathing plaster should be used and any cut ends of galvanized metal painted. Metallic conduit or channelling should also be treated to prevent rusting and provided with sufficient cover to prevent the risk of cracking the plaster-work.

If staining is known to emanate from clay bricks to be used as a background, specialist advice should be sought from the manufacturer. Sometimes an addition of lime to gypsum plaster is a satisfactory solution to the problem.

Mould growth. This may appear on newly decorated plasterwork due to persistently damp conditions. It can usually be remedied by removing the source of dampness, by improving the ventilation and allowing the structure to dry out completely. If the source of the damp conditions is condensation this may be due to the living pattern of the family in some dwellings. Existing mould and decoration should be scraped off the surface of the plasterwork and when dry, the infected area treated with a fungi-cide or bleach. This should destroy the spores produced by the mould and prevent the problem recurring.

Softness or chalky surface. This may be due to excessive suction of the undercoat, insufficient thickness of the finishing coat, continued working of the plaster after initial setting or because the final coat has been exposed to excessive heat or draughts during setting. Excessive suction of the background can be overcome by wetting or the use of a bonding agent.

Surface dampness. Recurrent surface dampness may be due to deliquescent salts attracting moisture from the air. The salts often emanate from unwashed sand used in the background which migrate into the plaster. It is advisable to use washed sands or other suitable sands complying with BS 1198–1200 (1976).

Popping or blowing. This can occur due to particles in the background or in the plaster expanding after the plastercoat has set.

8.2   Masonry lined with plasterboard

Dry and plastered wall linings can be used to improve fire resistance, thermal and sound insulation and provide water vapour resistance to reduce the risk of interstitial condensation.

Dry lining

Plasterboard used for dry lining is different from the plastering grade. The boards used for dry lining have an exposed ivory papered face for direct decoration and the back has a grey paper finish. The board edges can be supplied square, bevelled or tapered for smooth, seamless jointing. For increased thermal insulation an airspace at least 20 mm should be left behind the dry lining. The insulation value can be further increased by using insulation between supporting battens or by using foil-backed plasterboard.

Plasterboard can also be obtained with a water vapour-resisting polythene film bonded to the grey side and also with an integral backing of expanded polystyrene incorporating a polythene vapour membrane.

Handling. Boards should be off-loaded manually and carried on edge, two men per board. The boards should never be carried with their surfaces horizontal as this could impose undesirable strain on the plaster core. When stacking, the long edge should be placed down before being laid horizontal. Boards should never be dragged over each other as this can cause surface damage. The boards are supplied in pairs with the ivory-coloured surfaces together and bound along their cut edges. To protect the ivory surfaces the pairs should not be separated until required for use.

Storage. The boards should be stacked on a level surface in a dry storage area, preferably inside a building and protected from damp and the weather. If outdoor storage is unavoidable the stacks should be raised above the ground on a level platform, e.g. a suitable platform can be constructed from timber bearers 100 to 125 mm wide, not less than the width of the boards in length and spaced not more than 400 mm apart, at right angles to the boards. Outdoor stacks must be completely covered with a securely anchored polythene sheet or tarpaulin.

Methods of fixing

Plasterboard dry linings can be fixed on timber battens, plaster dabs or by 'dots and dabs'. There is also a proprietary system of screw fixing to metal channels which are also applied to the wall with a patent adhesive (British Gypsum White Book, 1986). The use of dabs of plaster for fixing the boards is only suitable for dry walls. It is undoubtedly the simplest method but true alignment of the boards requires considerable experience and skill. An alternative and more accurate method uses small pads or dots of bitumen-impregnated fibre board to provide a true, flat ground for the plasterboard. Dabs of plaster or adhesive are applied between the pads and the boards pressed on to them; when set, the dabs hold the boards in position. This system is normally used to secure 900 mm wide, 9.5 mm or 12.7 mm thick tapered edge boards to masonry backgrounds.

Fixing with timber battens may be used as an alternative to the above methods or specifically as a remedial measure to cure dampness. If the reason for this method of fixing is the latter the battens should be pressure-impregnated with preservative and fixed to the wall over strips of a damp proof membrane with compatible rust-resistant nails. The battens should have minimum dimensions of 50 mm width and 20 mm thickness and be fixed vertically at 400 mm or 600 mm centres for 1200 mm wide boards and 450 mm centres for 900 mm wide boards. Plasterboard is normally fixed to the walls vertically with the edges butt-jointed. All edges must be supported by battens or noggings. If walls are damp, remedial measures should be taken to cure the dampness before battens are fixed to the walls.

Another mechanical fixing method uses patent metal furring channels (British Gypsum White Book, 1986) for fixing tapered-edge plasterboard to masonry walls. The plasterboards are power-screwed to lightweight metal furring channels specially designed so they can be fixed vertically to wall surfaces with adhesive. All backgrounds to be dry lined using this method must be reasonably dry and protected from the weather. With backgrounds of smooth brick or some lightweight aggregate blocks it may be necessary to apply a bonding agent before use of the adhesive.

When applying the adhesive to low-suction backgrounds for fixing the metal sections the adhesive may slip. To avoid this it may be necessary to make a thin application of the material and to wait a short time before adding the thickness of adhesive required for bedding the channels.

With high-suction backgrounds bond failure of the adhesive may occur unless a polyvinyl acetate (pva) bonding coat is applied before application of the adhesive.

The lightweight metal channel sections are 50 mm wide by 9.5 mm deep, available in 2260 mm lengths and should be fixed to the walls at 600 mm centres vertically. Where horizontal services run behind the wall lining, the metal channels are cut to length and erected in two parts, leaving a gap sufficient to accommodate the service. This interruption in the vertical height may be up to 300 mm with the exception of cavity barriers and partition fixing points.

Jointing

Jointing of dry linings is made with tapered-edge plasterboard and carried out manually or mechanically using Ames mechanical jointing tools when there is a steady and reasonably large amount of work. With tapered-edge boards a flush seamless surface can be achieved. A bond of filler compound is applied to the trough of the tapered-edge joints and a suitable length of joint paper tape pressed into place with a filling knife. Finally a coat of filler is applied and finished smooth with a jointing sponge.

Cut edge joints are filled in a similar manner but the cut edges need to be chamfered with a 'Surform' type tool to an angle of about 45 degrees and approximately two-thirds the thickness of the board and the paper burr removed with sandpaper. Such joints are difficult to conceal and should as far as possible, be limited to internal angles.

Plastering on plasterboard

Most gypsum plasterboards are suitable for a skim coat or two-coat work if necessary but gypsum wallboard used for unplastered dry lining is a different grade and should not be plastered on the ivory paper-faced side. When single-coat work is used on plasterboard a 5 mm thickness of Class B, type b2 (retarded hemihydrate) final coat gypsum plaster should be used.

Mixtures containing lime or Portland cement are not suitable for direct application to the boards. Two-coat work is not normally necessary on plasterboard as it is generally possible to achieve a true and level surface. However, if two-coat work is required, an undercoat of sanded Class B, type a1, gypsum plaster (haired or fibred browning plaster). Any normal final coat can be used either neat or gauged with lime. Alternatively, two-coat work in premixed lightweight bonding (type a3) plaster without lime can be used. Lightweight browning plasters containing perlite are not recommended for this purpose.

A total plaster thickness of 10 mm is generally specified and the final coat should not be less than 2 mm. Mixes containing lime can take longer to dry out.

Drying out and maturing of plaster
For plaster to harden satisfactorily it is essential that the new work is adequately ventilated and an appropriate temperature maintained. If the temperature is too high the plaster may dry out too quickly but a little warmth helps it to harden. Conversely, if the work area is too cold the plaster may remain wet or condensation may occur. The continual presence of moisture will weaken the plaster. During cold weather newly plastered surfaces should be protected from frost and background heating provided.

Defects
Defects occasionally experienced when plastering on masonry backgrounds were discussed on pp. 162–65. The following specifically refer to plastered finishes on plasterboard:

Cracks following a definite line. Such cracks occur at joints or junctions in the plasterboard if the joints have not been correctly formed. When these cracks occur the joints should be cut out and re-filled using paper or gauze reinforcement.

Loss of adhesion. In single-coat work this usually indicates that the wrong type of plasterboard has been used, i.e. that intended for direct decoration. Other causes include dirty surfaces on the boards; the use of lime in the plaster or the wrong grade of plaster.

In two-coat work loss of adhesion may be due to weak undercoat or one which has not set when finished. Similarly, if an overstrong final coat has been used or an incompatible plaster.

The only cure for lack of adhesion is to strip the plaster and then: (a) check the suitability of the board for plastering; (b) if the backing is satisfactory treat the surface with one or two coats of a suitable polyvinyl acetate (pva) bonding agent and re-plaster; (c) in two-coat work, if the backing is satisfactory, strip off the final coat, allow the backing to dry out and roughen the surface, remove any loose material, apply a suitable pva bonding agent, and re-plaster.

Soft and powdery with fine cracks. This is usually due to rapid drying out before the plaster has set. Gypsum plaster requires as much water to make it set as is driven off from the gypsum during the manufacturing process. To avoid this problem: (a) use the correct grade of plaster and apply it in the recommended thickness: (b) when using a low wetting grade of finish, e.g. for single-coat work on plasterboard, wait for the advanced set before final trowelling; and (c) use Browning HSB plaster on solid high-suction backgrounds.

8.3   Rendering of external walls

Rendering of external walls may be carried out to provide weather protection either at the time of the initial construction or subsequently because of rain penetration problems, or for the sake of appearance. Lightweight renders can also be provided to enhance the thermal efficiency of an external wall and again this may be provided at the construction stage or later to upgrade an existing building.

The degree of resistance to damage by abrasion or impact that is required of a rendering varies with its location particularly in relation to adjacent footpaths and driveways, etc. This requirement affects both the strength of mix to be specified and the type of finish that is appropriate, since rough textured finishes are more liable to damage than smooth finishes.

Types of finish

Pebble dash or dry dash. This gives a rough finish of exposed pebbles or crushed stone graded from about 6 to 13 mm and is produced by throwing the aggregate on to a freshly applied coat of mortar. The aggregate is sometimes lightly pressed to and tapped into the mortar. A pebble dash finish on an undercoat with a spatterdash coat beneath it is particularly suitable for walls exposed for long periods to driving rain and wind and is frequently used in rural and coastal areas. It needs a strong but plastic mortar to secure the aggregate firmly and therefore should only be used on strong or moderately strong backgrounds.

Roughcast, wet-dash or (Scottish) harling
This is a rough finish produced by throwing on a wet mix containing a proportion of coarse aggregate. The aggregate in the finish coat is composed of sand and crushed stone or gravel from about 6 to 13 mm (perhaps even larger for harling) the proportions of sand and gravel being adjusted according to the effect required. Roughcast finishes on an undercoat with a spatterdash coat beneath it, like pebble dash finishes, are very satisfactory for use in severe conditions.

Smooth floated finishes
Surface crazing is always noticeable on smooth finishes and tends to produce a patchy appearance. The risk of cracking and crazing is greater with cement-rich mixes, and when fine sands are used. For smooth finishes a wood, felt or cork-faced float should be used, never a steel float. A coarse sand is recommended for this type of rendering (particularly where both rain and frost are likely to be severe) and walls should be well protected with a projecting eaves, etc.

Textured finishes
These are produced by treating freshly applied finishing coats with various tools to produce a variety of patterns and textures. Typical finishes are English cottage texture, town stucco, stippled stucco, fan texture and scraped finish.

The final coat of mortar is levelled and allowed to set for several hours and is then marked or scraped with the appropriate tool. Textured and scraped finishes are suitable for all backing materials and conditions of exposure and are less prone to crazing than smooth finishes. This is especially true of scraped finishes because the surface layer of very fine particles of sand and cement, which is likely to shrink and crack, is removed by the scraping action. In industrial areas the more heavily textured finishes may get dirtier than other types. A reasonably low-suction backing coat is recommended for highly textured finishes, otherwise the workability of the finishing coat tends to be lost before the required finish is achieved.

Machine-applied finishes such as Tyrolean have the final coat thrown or spattered on by a machine. The texture varies dependent upon the roughness of the material used and the type of machine. This type of finish is suitable for all walling materials and all conditions of exposure. Proprietary materials (some self coloured) are normally supplied ready for mixing, but if no special undercoat is supplied or specified by the manufacturer a 1:1:6 cement:lime:sand mortar is normally satisfactory.

Other finishes

Based on a variety of organic binders, these range from artificial stone paints and sand textured paints to coloured thick-textured compositions applied by spraying, usually by specialist contractors. Such finishes may be used as a decorative medium on a cement-based undercoat or applied directly to other suitable backgrounds such as certain types of blockwork. Cement-based undercoats should be 3–4 weeks old before applying organic finishes. The life of organic renderings is less than that of cement-based types and may need maintenance at intervals between 7 and 15 years.

Background characteristics

The characteristics of the background will, to a large extent, determine the type of undercoat that will need to be applied to it also the number of coats and the type of finish. Normal backgrounds come within the following categories and vary considerably in strength, suction rate and mechanical key:

High density. Very dense clay or concrete bricks and blocks and closed-surface lightweight-aggregate concrete. These units have low porosity and limited suction rate and frequently have smooth surfaces which provide little mechanical key and treatment to improve bond. Alternatively, raked out joints, bush hammering or fixing metal mesh to the surface may be adopted.

Moderately strong and porous. Most clay, calcium silicate or concrete bricks and blocks are in this category. These backgrounds normally offer some suction and mechanical key. Joints should normally be raked out unless the suction is irregular or too high in which case a spatterdash coat should be used. The latter may also assist in overcoming the effects of any soluble salts present in the units.

Moderately weak and porous. Materials such as open-surface, lightweight-aggregate concrete, aerated concrete and some relatively weak bricks come into this category. More care is needed in selecting renderings for such backgrounds as shrinkage of unduly strong renderings is liable to lead to shearing of the surface of the substrate. It is therefore important that the rendering is weaker than the background material.

Metal lathing. Expanded metal lathing or welded wire mesh backgrounds are frequently used for remedial work, particularly when the background material is friable. Ideally, stainless steel should be used but if ferrous metal is used it is essential to adopt three-coat work. A dense and relatively impervious first coat should be used to prevent rusting of the ferrous lathing and provide the necessary resistance to rain penetration.

High sulphate backgrounds. Some bricks and some old walls contain appreciable amounts of soluble salts, including sulphates. When saturated these salts can attack the tricalcium aluminate (C_3A) constituent present in all Portland cements. For further information see BRE Digest 89 (1971) and Chapters 1 and 6.

A good rendering will usually prevent rain penetration. However, if it is suspected that the background material is high in soluble sulphates the use of a rendering based on sulphate-resisting Portland cement is a wise precaution.

Walls damaged due to sulphate attack of the mortar joints should not be rendered.

Preparation of background

The background must be free of all dust, dirt, loose particles, laitance, old plaster, efflorescence, fungi or algae. Holes and depressions (other than raked out mortar joints) should be 'dubbed out' well before applying the undercoat. It is frequently necessary to dampen the background to reduce excessive suction and to provide an even suction over the area to be rendered.

Bonding agents, usually based on polyvinyl acetate (pva) emulsion, may be used to provide a key but there is no substitute for a good mechanical key.

Choice of mix

Renders should generally be porous and of low strength in order to reduce the risk of large cracks occurring due to drying shrinkage during the initial drying out period. Such porous rendering will have an 'overcoat effect', i.e. it will absorb some rain-water but will not readily pass it on to the background and it will quickly dry out during the next fine spell. Weak porous renders are usually adequate in all but the most exposed areas. Walls exposed to driving rain and wind will need at least one coat to be fairly impervious, this coat should always be the first.

Frost damage to rendering will only occur if the material is thoroughly saturated and is unlikely if the walls are well protected by wide eaves and good detailing to exclude rainwater. In very exposed situations where the walls are likely to be exposed to hard frosts and/or continuous driving rain the use of pebble dash, roughcast, wet-dash or (Scottish) harling is recommended as these finishes shed the rainwater.

Waterproofing admixtures can be used but are not generally recommended for use in intermediate coats as they may be responsible for a loss of bond. If used, the manufacturer's recommendations should always be strictly complied with.

Choice of mix will therefore depend on: (a) the background material; (b) the exposure of the wall; and (c) aesthetic considerations.

Mixes

Table 8.4 gives mixes suitable for rendering and Table 8.5 shows the recommended mixes in relation to the background, the type of finish required and the degree of exposure.

Spatterdash mixes
These are used to provide a key on smooth backgrounds or as a relatively imperme-able first coat in conditions of severe exposure and consequently should be related to the suction of the background. Mixes of 1:2 cement:clean coarse sand mixed with

TABLE 8.4 Mixes suitable for rendering

Mix designation	Cement:lime:sand	Cement:ready-mixed lime:sand		Cement:sand (using plasticizer)
		Ready-mixed lime:sand	Cement:ready-mixed material	
(i)	1:¼:3	1:12	1:3	–
(ii)	1:1½:4 to 4½	1:8 to 9	1:4 to 4½	1:3 to 4
(iii)	1:1:5 to 6	1:6	1:5 to 6	1:5 to 6
(iv)	1:2:8 to 9	1:4½	1:8 to 9	1:7 to 8

When soluble salts in the background are likely to cause problems, mixes based on sulphate-resisting Portland cement should be used.

TABLE 8.5 Mixes for external renderings based on background materials, exposure conditions and finish

Background material	Type of finish	First and subsequent undercoats			Finishing coat		
		Exposure			Exposure		
		Severe	Moderate	Sheltered	Severe	Moderate	Sheltered
High density Strong	Wood float	ii or iii	ii or iii	ii or iii	iii	iii or iv	iii or iv
	Scraped or textured	ii or iii	ii or iii	ii or iii	iii	iii or iv	iii or iv
Smooth	Roughcast	i or ii	i or ii	i or ii	ii	ii	ii
	Dry dash	i or ii	i or ii	i or ii	ii	ii	ii
Moderately strong and porous	Wood float	ii or iii	iii or iv	iii or iv	iii	iii or iv	iii or iv
	Scraped or textured	iii	iii or iv	iii or iv	iii	iii or iv	iii or iv
	Roughcast	ii	ii	ii	ii	ii	ii
	Dry dash	ii	ii	ii	ii	ii	ii
Moderately weak and porous*	Wood float	iii	iii or iv	iii or iv	iii	iii or iv	iii or iv
	Scraped or textured	iii	iii or iv	iii or iv	iii	iii or iv	iii or iv
	Dry dash	iii	iii	iii	iii	iii	iii

* Finishes such as roughcast and dry dash require strong mixes and are not recommended on weak backgrounds.

sufficient water to give a thick slurry are suitable if there is some suction and should be dashed on the wall with a trowel or scoop to give a thin coating with a rough texture. This coat should be kept damp initially and then allowed to dry slowly before the undercoat is applied. If negligible suction is available a mix of 1½:2 cement:sand can be used with an addition of pva emulsion as recommended by the manufacturers. The slurry should be applied to the surface and stippled with a brush to form a deep, close-textured key. This type of spatterdash coat should be allowed to dry slowly and harden before the undercoat is applied.

Sand

The choice of sand is extremely important as it affects the working properties of the mix, its water demand and the performance and appearance of the finished rendering. The sand should be well graded to BS 1199. Aerated rendering mixes are particularly sensitive to the grading of the sand and to the suction of the background. The suction of the background removes water from the mix and tends to break down the bubble structure during subsequent working, particularly at the interface with the background material. Sand for aerated mixes should therefore contain sufficient fine material to help stabilize the bubble structure and maintain some workability if the air content is reduced.

Undercoats require the coarsest and sharpest sand that can be handled but the most suitable grading of sand for the final coat will depend upon the finishing treatment. For floated, scraped or dry-dash finishes the same grading as used in the undercoat will be suitable but it may be desirable to have less coarse material for those textured finishes produced by treatment of the freshly applied final coat with a tool. For torn textures a slightly higher proportion of material larger than 5 mm is desirable.

Undercoats
The undercoat should always be weaker than its background. In general the same mix should be used for the finishing coat as for the undercoat but if it is necessary to depart from this procedure the undercoat should be stronger than the finishing coat, never the reverse. A rich finishing coat over a lean undercoat can cause serious cracking and loss of adhesion. The undercoat should dry out thoroughly before the next coat to accommodate maximum shrinkage. It is recommended that at least 2 days be allowed for drying of the undercoat in summer and 1 week or more in cold or wet weather. It is essential that spatterdash coats and all areas of dubbing out have hardened before the undercoat is applied.

Thickness and number of coats

One-coat work is sometimes used but is not recommended. External renderings should consist of not less than two coats. One undercoat and a finishing coat are normally adequate but extra coats are necessary on metal lathing to level the uneven surface. Spatterdash coats are normally 3 to 5 mm, undercoats 8 to 16 mm and final coats 8 to 10 mm but some finer textured machine-applied finishes may be as thin as 3 mm.

In conditions of severe exposure three-coat work is recommended and for moderate conditions two-coat work. Expanded metal sections should be used to provide neat and protected finishes at edges, angles and stops.

Exposure conditions

The terms severe, moderate and sheltered exposure used in Table 8.5 are based on the definitions in BRE Digest 127 (1971) (see also Chapter 7) and are derived from driving rain indices (BRE Report, *Driving rain index*, 1976) as modified by local conditions and building heights.

Reinforced renders

Rendering reinforced with alkali-resistant glass fibres tend to prevent the formation of drying shrinkage cracks in the rendering. This special type of rendering is normally applied as a single 5 to 8 mm coat and its overall moisture movements are

therefore more easily restrained, even by weak backgrounds, than those of traditional cement-rich renderings (BRE Digest 196, 1976).

Detailing

BS 5262 (1976) gives advice of detailing of sills, string courses, parapets and other
architectural features to minimize streaking and to improve the performance of
rendering.

External insulation

Lightweight insulating renders are now being used and are typically cement-based
and incorporate expanded polystyrene or a similar material as the aggregate.
Applied densities generally range from 300 to 400 kg/m^3. Lightweight insulating
renders are less thermally efficient than most other forms of added insulation but
do have advantages in overcoming detailing problems (Roberts, 1980). The substrate
needs to be prepared to receive the render in a similar manner to traditional render.
Usually renders of this type need to be protected with a comparatively dense decorative finish.

Coloured renders

Colouring of renders can be achieved using one of the following methods:

(a)   Using a suitable sand or coloured aggregate.
(b)   Using white or coloured Portland cement.
(c)   Incorporating pigments complying with BS 1014 (1975) into the render mix. It
      can be difficult to achieve true colour consistency. Additions of pigment should
      be strictly in accordance with the manufacturer's recommendations.
(d)   By using a ready-mixed lime–sand rendering which contains a controlled quan-
      tity of pigment. Dark colours are not recommended as any defect such as
      efflorescence, lime bloom or pattern staining can badly disfigure the wall finish.
(e)   Using proprietary finishing materials.

8.4   Painting masonry walls

If painting of newly constructed walls is delayed until they are completely dry many
of the common defects associated with paint would not arise. The main causes of
trouble are:

(a)   Moisture in the walls due to the building process, water penetration because of
      the construction defects or condensation
(b)   Variations in texture and porosity of the wall finish
(c)   Soluble salts producing efflorescence and alkalis from the materials used in
      construction and/or plastering
(d)   Unsuitable conditions inside the building at the time of painting
(e)   A defective substrate for the paintwork, e.g. faulty plasterwork
(f)   The wrong choice of paint for a particular background, whether new or existing.

TABLE 8.6 Choice of paint system (some examples)

Type of paint	General comments	External durability	Alkali resistance
Cement paints*	Provide a hard, durable semi-rough surface but soils quickly and erodes rapidly in highly acid atmospheres; prone to algae growth in wet areas. They are unsuitable on gypsum plaster. Cement paints shold not be used on any background containing water-soluble sulphates. Sound coatings can be painted over with alkali-resisting primer and oil paints, chlorinated rubber paint or emulsion paint.	5–7 years	High
Emulsion paint, general purpose good quality*	Some types are unsuitable in kitchens and bathrooms. Fairly good wear resistance and washability. Can be repainted with emulsion or oil-based paint.	3 years	Low
Emulsion paint, matt or contract lower quality*	As for general purpose good quality emulsion paint but inferior washability.	Unsuitable	Medium
Emulsion paint exterior smooth*	Exterior grades based on acrylic, styrene/acrylic and pva/verstate copolymers.	4–7 years	Should be high
Emulsion paint, exterior sand-textured or fibre filled*	These paints have thicker films than other emulsion paints. External durability depends to some extent on film thickness; fibres are not always beneficial from this point of view.	4–10 years depending on thickness	High
Emulsion paint (vinyl) silk, satin or gloss*	Opacity decreases as gloss increases; silk and satin said to have good resistance to washing. Vinyl appears to relate to quality rather than composition.	Unsuitable	Low or medium
Textured 'plastic' paint*	Some plastic paints such as Artex are fairly porous unless a glaze is used. Some are only washable if a primer and/or a paint system is applied.	Unsuitable	Low
Textured or coarse mineral aggregate filled emulsion coatings*	Variety of textures, applied by brush, roller or trowel. Applied in thicknesses up to 2 mm.	10 years	High
Gloss paint (mainly alkyd resin based)+	Most good quality paints are suitable for use outside. Some cheaper paints are only suitable for internal use.	3–6 years	Low
Polyurethane one pack)+	As for gloss paint.	3–4 years	Medium
Exterior masonry paints (smooth, matt or semi-gloss)+	Based on isomerized or synthetic rubber, vinyl or 'acrylic' resins.	4–7 years	High
Chlorinated rubber (masonry) paints+	For special purposes only, e.g. water, acid or alkali resistance. These paints are not resistant to oils and solvents. They have a moderate gloss.	4–7 years	Very high
Epoxy ester paints (one pack)+	Better water and chemical resistance than hard gloss paints. Sometimes used as primers.	3–5 years	High

* Water-thinned; + Solvent-thinned.

When it is not practical to leave the wall surface until it has thoroughly dried out, temporary porous finishes such as emulsion paint, can be applied (or cement paint on renderings) as soon as the surface is dry and efflorescence (if any) has ceased to form. Such finishes will let the wall continue drying out, whereas gloss and similar more impervious finishes are liable to restrict drying out. Problems are likely to occur when incompletely dried walls are sealed on one or more surfaces with relatively impervious materials.

When partially dry walls are decorated with porous finishes drying out can be encouraged by gentle heat and ventilation, but heaters burning liquid or gas fuels should not be used as these tend to induce additional moisture into the atmosphere.

Porous paints used for temporary decoration should be compatible with the proposed permanent decoration or be easily removable.

Characteristics of wall surfaces

Lime plasters

Traditional non-hydraulic plasters are rarely used except for restoration work and the usual decoration is a limewash, details of such finishes are given elsewhere (Ashurst and Ashurst, 1988) although emulsion paints can be used after an extended drying out period (BRE Digest 197, 1977). These plasters harden slowly on ageing and sometimes develop shrinkage cracks. They are free from soluble salts and caustic alkali. Hydraulic lime is rarely used for finishing plaster and has a variable caustic alkali content which can attack paint severely.

Gypsum plasters

When used neat, in the fully set, dry state, they can be painted without difficulty using almost any type of paint (BRE Digest 197, 1977). They have a hard, smooth surface, the porosity varying dependent upon the class of plaster used. Adhesion of paint is usually good, even on the denser plasters unless they are trowelled to a glazed surface, but is seriously reduced by dampness, particularly on the denser types.

Additions of lime, or cement in the backing coats, may be responsible for the development of alkalinity and necessitate use of special primers or paints.

BS 1191: Part 1: 1973 quotes two classes of gypsum plaster as follows:

Class A Plaster of Paris and Class B retarded hemi-hydrate (e.g. Thistle); Class A plasters should be treated as Class B for painting but as they are generally only used for repair work and may create a variable suction, it is normally recommended that a sealer be used before painting.

Class B plasters have a smooth, hard, but moderately porous surface to which paint adhesion is good, but absorption of water-thinned paints is sometimes patchy. The alkali content is small and quickly reduces on ageing; however, lime may be added but this accelerates the set and alkalinity. Over-wetting during plastering or continuously damp conditions (due to water penetration or condensation) may cause 'sweat out' leading to softening and subsequent paint failure.

Lightweight gypsum plasters

BS 1191: Part 2 (1973) specifies two types of pre-mixed lightweight plasters: (a) undercoat plasters (Carlite-Browning; Browning HSB; Bonding coat; Welterweight

Bonding coat and Metal Lathing); and (b) Final-coat plasters (Carlite Finish and Limelite Finish).

These plasters are for general use, especially when it is desired to increase fire resistance, marginally improve thermal insulation or reduce surface condensation. If the latter two properties are required, impervious paints should not be used.

Lightweight gypsum plasters can usually be treated as for Class B plasters, being based on retarded hemi-hydrate plasters with added perlite or vermiculite. They may have high or variable suction and take longer to dry out than Class B plasters. According to the BRE (British Gypsum White Book, 1986) on a background liable to produce stains or efflorescence, an aerated sand/cement backing is desirable. Also patchy absorption of emulsion paint can result from uneven drying out and successive coats may 'pile up' on the more absorptive areas rather than even them out. A thin, non-aqueous paint or sealer is necessary to overcome this difficulty.

Thin-wall plasters

These are based on mineral fillers and textured finishes such as 'Artex' and offer no restriction to early decoration as they dry out rapidly. They are usually fairly porous and may need a sealer coat.

Portland cement backgrounds

Renderings, mortar joints and cement based units, etc. are nearly always sufficiently alkaline to attack paint under damp conditions. In addition, alkalinity from Portland cement-based undercoats can pass into a gypsum plaster final coat and create similar problems.

Free lime usually present in cement plasters is a cause of lime bloom which is particularly noticeable on emulsion paints. If carbonation is allowed to occur before painting it can greatly reduce this effect and also reduce surface porosity.

Brick and stone

The porosity/absorption varies depending upon the type of unit from almost zero absorbency for some igneous stones and highly vitrified bricks to quite high values for calcium silicate bricks and Flettons. Alkalinity is not usually significant (with the exception of the mortar joints) but soluble salts are present in some bricks and may cause efflorescence. The low-absorption surfaces require strongly adherent paints such as chlorinated rubber and epoxy or polyurethane formulations. The more highly absorbent units may have marked variation in suction rate from one unit to the next, necessitating thin, penetrating first coats. Emulsion paints differ in their adhesion to some high-suction bricks and specifiers should consult the manufacturers of the units for advice if in doubt. Adhesion of all paints is much better on sand-faced bricks.

New and re-pointed masonry will require alkali-resistant paints because of the mortar joints.

8.5   Tinting

Over the past 35 years numerous methods of tinting defective masonry have been used with varying degrees of success and even using modern tinters failures can occur due to the following:

(a)   the solution being applied to wet or damp masonry;
(b)   the solution being applied to a low absorption unit (i.e. below 8–10 per cent) unless a surface treatment is relied upon;
(c)   the method of application being incorrect for the units concerned, i.e. the tinter is applied too thickly or too thinly.

Generally tinters have a low viscosity and are absorbed into the units or mortar thus allowing the masonry to perform in a similar manner to the untreated surfaces. This is important as to alter the surface characteristics of the wall can cause difficulties, particularly if only part of the surface has been treated and differing moisture movements occur.

Tinters are usually a suspension of insoluble mineral/synthetic pigments (generally metallic oxides) in a clear liquid base. Manufacturers usually claim that the colouring agents are resistant to ultraviolet light and weak acids and comply with BS 1014 (1975). Manufacturers also claim that when applied their low viscosity allows penetration into the substrate and the formulation of the liquid base ensures adhesion of the pigments.

For less absorbent smooth-faced units some manufacturers use a range of acrylic resin-based tinters which perform in a different manner to the more readily absorbed low viscosity tinters and almost certainly alter the surface characteristics of the units treated.

Mortar tinting to overcome colour variations is often carried out but it is important to bear in mind that mortar, being a relatively porous cementitious material, is liable to erode due to the acidity of rainwater or frost action. This results in the gradual loss of the surface from the joints and consequently any tinter which may have been applied.

Most tinting contractors guarantee their work but frequently exclude from the warranty the following:

(a)   work carried out during unsuitable weather conditions;
(b)   work carried out on a damp or wet surface;
(c)   work below d.p.c. level;
(d)   surfaces previously treated with any surface coating or liquid;
(e)   any other parties' workmanship;
(f)   failure of the substrate materials;
(g)   any atmospheric effects, e.g. acid rain, etc.
(h)   variable weathering of the wall surfaces.

Tinting of masonry is frequently successful on the appropriate masonry materials under controlled conditions but specifiers are advised to carry out small site tests in inconspicuous locations before deciding upon whether or not to proceed.

References

Ashurst, J. and Ashurst, N. (1988) *Practical building conservation. English Heritage Technical Handbook Vol 3, Mortars, plasters renders.* Gower Technical Press.

BS 1191: Part 2: 1973 *Gypsum building plasters.* BSI, London.

BS 1198–1200 *Building sands from natural sources.* BSI, London.

BS 5492: 1977 *Code of practice for internal plastering.* BSI, London.

BS 5262: 1976 *Code of practice for external finishes.* BSI, London.

BS 1014: 1975 *Pigments for Portland cement and Portland cement products.* BSI, London.

BS 4049: 1966 *Glossary of terms applicable to internal plastering, external rendering and floor screeding.* BSI, London.

British Gypsum White Book (1986) *Technical manual of building products* British Gypsum.

Buildings Research Establishment Digest 49 (1964) *Choosing specifications for plastering.* Buildings Research Establishment.

Buildings Research Establishment Digest 89 *Sulphate attack on brickwork.* Buildings Research Establishment.

Buildings Research Establishment Digest 127 *An index of exposure to driving rain.* Buildings Research Establishment.

Buildings Research Establishment Digest 196 *External rendered finishes.* Buildings Research Establishment.

Buildings Research Establishment Digest 197 (1977) *Painting Walls Part 1: choice of paint.* HMSO.

Buildings Research Establishment Report *Driving rain index 1976.* Buildings Research Establishment.

Roberts, J. J. (1980) *External insulation.* C and CA Reprint 4/80.

9

Thermal and sound insulation of walls

9.1 Thermal insulation

Building Regulations relating to energy efficiency have become increasingly more stringent over the years but the aim of the Regulations in respect of external walls has always been 'to limit the heat loss through the fabric of the building'. The method of theoretically evaluating the heat loss through external walls is by calculating the thermal transmittance coefficient or U value of the construction, i.e. the rate of heat transfer in watts through 1 m^2 of a wall when the temperature on each side of the wall differs by 1°C (expressed in W/m^{2}K).

The current Building Regulations (England and Wales) specify a U value of 0.45 W/m^{2}K for walls and floors when compliance is being sought using the Elemental Method. Alternatively, compliance with the Building Regulations can be demonstrated using Target U value or Calculation Procedure methods. These methods allow higher U values at some parts of the fabric (i.e. values higher than specified in the Elemental Method) to be compensated for by lower U values elsewhere. The following calculations illustrate how a variety of alternative arrangements of masonry and masonry plus added insulation can meet U values of 0.6 or 0.45 W/m^{2}K and the same calculation methods can be used to determine any other U value.

Improved wall insulation can be achieved by a variety of alternative arrangements of masonry and masonry plus added insulation as follows:

(a) Cavity walls with clear cavity. Using standard thickness external leaves (usually 100 to 103 mm) and increasing the lightweight block thickness above 100 mm.
(b) Cavity walls with partially insulated or totally insulated cavities.
(c) Solid walls with or without added insulation.

Example 1 – cavity wall with clear cavity
Figure 9.1 illustrates a construction consisting of 102.5 mm brick, 50 mm cavity, 125 mm block (density 475 kg/m^3), 13 mm lightweight plaster giving a U value of
180

The use of cavity fill, particularly total fill, increases the risk of rain penetration especially if a clean cavity is not maintained, particularly in areas of high exposure.

If full cavity fill is used in exposed areas it may be advisable to consider rendering the external face. In addition, BRE data suggest that certain insulation materials have better resistance to rain penetration, i.e. polystyrene beads and blown-in rock fibre. Using a wider cavity lessens the possibility of its becoming filled with building debris and of course enables a better insulation value to be achieved. However, wet insulation does not improve the thermal properties of a wall and may be responsible for subsequent problems.

It should also be remembered that due to the introduction of full cavity fill the outer leaf will be cooler and possibly wetter and some masonry materials may be more prone to frost/sulphate attack as a result of these conditions. Also some insulants may cause weep holes to become ineffective and there may be a danger of a reservoir effect at cavity trays over lintels, etc.

Example 4 – solid walls
Figure 9.4 illustrates one method of achieving a U value of less than 0.6 W/m²K using 190 mm lightweight blockwork (650 Kg/m³) with 16 mm render externally and 9.5 foil-backed plasterboard on battens internally. This construction gives a U value of 0.52 W/m²K.

Values of 0.35 W/m²K or less can similarly be achieved by using external insulation rendered, a thicker block and foil-backed plasterboard on battens internally.

Solid walls used in the past have sometimes had problems of dampness and rain penetration, but with modern materials and techniques improvements on old methods can easily be achieved and using external insulation there can be advantages:

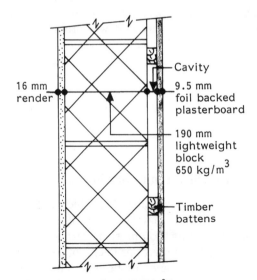

16 mm render

Cavity

9.5 mm foil backed plasterboard

190 mm lightweight block 650 kg/m³

Timber battens

Figure 9.4   U value 0.52 W/m²K

(i)   The thermal capacity of the masonry is used (i.e. the masonry acts as a storage heater) and improves the comfort of the occupants. There does not appear to be any evidence that this saves energy but the gradual warming up of the construction and the delayed cooling provides a more comfortable environment unless the temperature of the structure is allowed to fall too low.
(ii)   In consequence of (i) potential interstitial condensation problems are largely avoided.
(iii)   Cold bridging problems are avoided.
(iv)   There is no loss of internal space.
(v)   The acoustic performance is better.
(vi)   The insulation and masonry are kept dry and should provide improved performance.

The disadvantages are:

(i)   Limitations on the type of finish and appearance.
(ii)   The vulnerability of some systems to vandalism.
(iii)   Problems of detail at openings in existing construction.

Special units

In addition to the traditional solid, cellular and hollow concrete blocks, a wide variety of dense and lightweight concrete sandwich panels, foam-cored units and solid panels with insulated dry lining are currently available and no doubt in view of the likely requirements for higher thermal performance walling many more will appear in the future.

Double leaf block walls
A variety of alternative double leaf block walls with 50 mm cavity can achieve a U value of 0.6 W/m^{2}K or better, e.g. using lightweight blocks a 140 mm inner leaf with lightweight plaster, 50 mm cavity, 100 mm outer leaf rendered will produce a U value typically between 0.53 and 0.63 W/m^{2}K. A further improvement can be achieved if the lightweight plaster is replaced with plasterboard on battens, giving U values between 0.45 and 0.52 W/m^{2}K.

Similarly, if the external rendering is replaced by tile hanging or other forms of cladding such as weatherboarding, U values of 0.45 W/m^{2}K and better can be achieved.

9.2   Assessing masonry walls for cavity fill thermal insulation

External cavity wall construction was introduced to overcome the problems of rain penetration through external walls. It therefore follows that if cavity walls are to be filled with cavity insulation material, detailed investigation must be carried out to ensure that the walls are suitable for this treatment to prevent problems of dampness occurring. It also follows that if certain insulating materials become damp their insulating properties are nullified.

Many designers (including the writer) are opposed to completely filling cavities. However, full cavity insulation is permissible under certain circumstances and the purpose of this section is to discuss assessment of the suitability of external cavity walls for filling with thermal insulants.

BS 8208: Part 1: 1985 gives detailed guidance on the necessary assessment required before the use of total cavity fill and would-be users are strongly recommended to study that document. It is assumed that assessment of cavity walls will be carried out by appropriately qualified and experienced people and that if the walls are considered suitable for filling with insulants that the work will be carried out by well-trained installation technicians.

Factors to be considered

Form of construction
The form of construction must be two leaves of masonry with a cavity of at least 40 mm between the two leaves. Timber frame construction with an external veneer of masonry is not suitable for filling with insulation between the external leaf of masonry and the inner timber frame leaf. Walls having a height in excess of 12 m are not generally considered suitable for totally filling with insulants. Some forms of construction having structural frames may be suitable (e.g. reinforced concrete) provided that the frames do not bridge the cavity.

Site conditions
The exposure of wind-driven rain needs to be assessed and any problem known to have arisen due to the exposure should be noted in relation to the form of construction used.

If problems of dampness have arisen in the past these may influence the assessor's decision as to whether total cavity fill is appropriate. This will undoubtedly depend to some extent upon what the problems were and how they were overcome.

Where cracks exceeding 1 mm in width are noted in the external leaf of masonry expert advice should be sought.

Special attention should be paid to buildings where less than 3 years have elapsed since the first occupancy, as defects in the construction may not have been exposed to the worst weather conditions for that site during such a short period of time. Equally, other defects such as settlement may not have become apparent.

Whenever possible, original drawings should be checked to ensure that the construction is appropriate.

Conditions of cavity
The cavity should be examined using a Borescope to ensure that it is free from obstructions bridging the space between the inner and outer leaves. Unacceptable bridging of the cavity may be caused by the following:

(a)   wall ties covered in mortar droppings
(b)   debris such as broken units or mortar/concrete
(c)   brick, stone or timber ties
(d)   snapped headers used in bonds other than stretcher bond

(e)   natural stone or no-fines concrete which may create a rough surface within the cavity.

Special care may need to be exercised to ensure that the underside of features such as cills and floor slabs do not act as a bridge and allow rainwater to penetrate the cavity.

Extent of cavity to be filled
It may be necessary to seal the cavity in certain areas to avoid filling unnecessary voids, e.g. if the cavity extends significantly (more than 0.5 m) below ground level or in a basement wall and procedures to prevent cavity fill entering the cavity below external ground level may need to be specified.

Vertical closers may be required to prevent fill entering the cavity where it extends beyond the limits to be filled, e.g. semi-detached houses, etc. where only one dwelling is to be insulated. It is essential that the head of the cavity is sealed to avoid displacement of loose fill or filling of voids if foam is used.

Outer leaf
Where the masonry is protected by cladding this should be watertight and gaps between the cladding and masonry at openings or the periphery may need to be sealed.

If the masonry is rendered it must be in good condition and free from sulphate attack or drying shrinkage problems (see BRE Digest 75; 1976). Where a paint of very low permeability has been used, e.g. chlorinated rubber, the wall may be unsuitable for filling and specialist advice should be sought.

The outer leaf of the masonry should be checked for cracks, gaps, holes, spalled masonry and defective pointing. If cracking is detected the cause should be ascertained. Horizontal cracking along separate mortar joints may indicate corroded vertical twist type wall ties, while other crack patterns may be due to movement or sulphate attack. Where cracking and/or spalling of units is extensive or the cause of cracking is uncertain specialist advice should be sought. Gaps around windows and doors and unsealed movement joints should be sealed, see BS 6213 (1982). Unfilled perpends (other than weepholes) should be made good, also defective pointing.

Some common features may cause local concentrations of rainwater on the wall, e.g. flush cills or copings/cappings without drips; impermeable cladding or glazing causing runoff onto the wall; lack of suitable overhangs; faulty chimney details; string courses; faulty gutters and downpipes also incorrectly installed overflow pipes.

Inner leaf
If the wall is plastered it must be in good condition and signs of dampness investigated, i.e. whether the dampness is due to water penetration, rising damp or condensation.

The type of masonry should be checked and the condition recorded, as well as its condition and holes for pipes, electric cables, etc. It is particularly important that all

holes and gaps are sealed to avoid the ingress of fill and to limit any gas
the fill from entering the living accommodation. For the same reason a ch
be made to note whether there are any discontinuities in the inner leaf, p.
adjacent to suspended ceilings, boilers, cupboards and cloakrooms, etc.

Dry lining on a permeable masonry inner leaf may lead to the ingress of
evolved by the fill into the building, depending on the particular cavity insulation
system.

Services in cavity

Services should not generally be permitted in cavities. The location of any pipes
should be determined, e.g. water, gas and oil since they could be damaged during
the installation process, particularly during drilling.

Similarly, electric cables should be located and may need isolating to protect them
from certain types of insulation material, e.g. the plastic sheathing may be damaged
if in contact with polystyrene. Enclosing of cables by fill can also cause their temperature to rise; hence it is advisable to check the power requirements for electric
cookers, etc. before installation of any cavity fill.

Flues bridging the cavity may also cause problems following filling of the cavity.

Ventilation

A check should be made to ensure that all vents remain clear, e.g. vents for appliance
air supply, fume extraction, necessary room ventilation, larders, ventilation of the
structure, in particular the sub-floor and roof voids.

Dampness risks

In a joint report (1983) the **BBA**, **BRE**, **BEC** and **NHBC** spelt out the risks of
dampness and how to minimize them following insulation of cavity walls. The report
confined its comments to walls having an inner leaf of blockwork and to four types
of cavity insulation; urea–formaldehyde foam; blown-in fibres and granules; mineral
(rock) wood or glass fibre slabs; rigid and semi-rigid insulation boards.

Based on their collective experience the report stresses:

(1)   The risk of rain penetration across a cavity wall, filled or unfilled, depends
       mainly on the amount of water penetrating the outer leaf. It is determined both
       by exposure of the wall to driving rain and the material and workmanship of
       the brickwork.
(2)   Brickwork in any exposure condition is less likely to suffer rain penetration if
       an absorptive rather than a dense brick is used.
(3)   Recessed joints significantly increase the chances of rain getting through an
       outer leaf.
(4)   Unfilled cavity walls may perform satisfactorily with minor imperfections in
       workmanship. When filled or partly filled, these imperfections assume greater
       significance and there is increased risk of dampness.

9.3 Sound insulation of party walls

Noise has been described as unwanted sound and in addition to causing annoyance, may in certain circumstances, induce stress, damage hearing or disturb concentration, thus affecting an individual's well-being and working efficiency (BS 2750: Parts 1 and 4: 1980).

The best defence against noise, both internally and externally generated, lies in the reduction of noise at source, good planning and design although in some cases no design, however careful, can eliminate problems due to bad planning.

This section is concerned with walls which separate habitable rooms within a dwelling from another part of the same dwelling which is not used exclusively by one family, i.e. Part E of the Building Regulations relating to Party Walls which states that they shall have 'reasonable resistance to airborne sound'.

There are two sound sources which may be discernible in a neighbouring dwelling: (a) airborne sources such as speech, musical instruments and loudspeakers; and (b) impact sources such as footsteps and the moving of furniture. Whatever the source, good sound insulation reduces the nuisance value but different techniques are often required to combat the two sound sources. This section is concerned only with airborne sound.

The structure

Sound insulation depends on the building as a whole and as much noise can be transmitted through flanking walls and floors (and sometimes the roof) as passes directly through a wall itself. Noise may be leaked through a crack or underfloor airspace where these are linked by gaps around joists built into the wall. Similarly, in roof spaces unfilled perpend joints in masonry may be responsible for leaked noise and frequently the materials of construction are changed above ceiling level, resulting in varying standards of workmanship and sound insulation.

Generally, designs which minimize common wall area are beneficial for sound insulation. BRE (Digest 252, 1981) suggest for example, that a wall will provide better sound insulation where it separates two rectangular rooms by forming one of the shorter dimensions rather than one of the longer dimensions. Similarly, stepped or staggered layouts between dwellings tend to improve performance. This applies mainly to airborne sounds, such as conversation or music. However, occupants are often disturbed by 'structure-borne' sounds such as plumbing noises, etc. It is generally preferable to arrange for rooms of similar function to be juxtaposed at the party wall and be vertically in line in blocks or flats.

Measurement of sound transmission

Acoustics is a branch of science for which a number of simple laws have been formulated. However, these laws are frequently applied to inappropriate forms of construction. The classic example is the 'mass law' which serves to give a theoretical indication of sound insulation. The mass law assumes: (a) a homogeneous wall; (b) a specific part of the frequency spectrum; and (c) ideal conditions. In practice all walls are non-homogeneous, it is necessary to consider the whole frequency range and not the region controlled by the mass law, conditions are never ideal and flanking transmission is the norm rather than the exception. Nevertheless, weight and mass are accepted as a criteria for determining an 'indication' of sound insulation for some materials and forms of construction, i.e. dense masonry.

As laboratory testing is usually carried out on a section of a wall attached at its edges to a heavy, rigid frame which is not representative of the conditions in a completed dwelling and therefore does not take flanking transmission into account, such testing can have only limited application. Even when field tests are carried out there is likely to be variability in results in different buildings on walls or floors of nominally the same construction. Hence, in order to assess the overall performance of construction against performance standards, a number of field tests would be required on different sites. The greater the variability the larger the number of tests which will be needed to provide an accurate assessment.

Field tests tend to give poorer results than laboratory results and even with field tests varying results can be achieved if the rooms either side of the test wall have a different geometry dependent upon which room is used as the sound source.

Unfortunately there is rarely a true correlation between laboratory and field tests.

Satisfactory wall construction

Solid dense masonry walls depend upon their weight to reduce vibration whereas cavity walls (dense or lightweight) with two or three leaves depend partly on their weight and partly on structural isolation between the leaves. The weight of masonry is generally accepted as the primary factor but stiffness and damping are also important. Consequently, walls of the same type but made from materials with different physical properties may need different weights to give the same sound insulation.

Solid walls

Example 1. Figure 9.5 illustrates a brickwork wall with 12.5 mm plaster either side which satisfies the Building Regulations (1985) Part E.1. If deep-frogged bricks are used the bricks should be laid frog up and the frogs filled with mortar. The brickwork bond should include headers (e.g. English or Flemish bond). The weight of the wall (including the plaster) should be at least 375 kg/m^2. All mortar joints should be carefully filled and a seal provided between the wall and other parts of the construction to achieve the specified weight and to avoid air leakage.

Example 2. Figure 9.6 is similar to example 1 but using plasterboard 12.5 mm, with each face secured with plaster dabs or timber battens. The brickwork should not be rendered or plastered to enable the sound absorbency of the brickwork faces to come into play and avoid the danger of drumming.

Example 3. Figure 9.7 is similar to example 1 but using concrete blockwork with a weight of 415 kg/m^2 (including plaster). The blocks should extend to the full thickness of the wall.

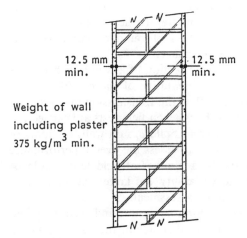

Figure 9.5   Sound insulation – brickwork plastered on both faces

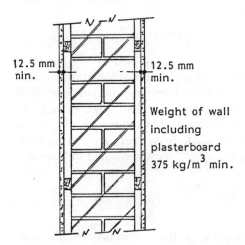

Figure 9.6   Sound insulation – brickwork with plasterboard on both faces

Cavity walls (single cavity)

Example 4. Figure 9.8 illustrates two leaves of brickwork or dense blockwork with 12.5 mm plaster on each of the room faces. The width of the cavity should be at least 50 mm and the weight of the wall should be at least 415 kg/m^2 (including the weight of the plaster). Construction requirements as for example 1.

Example 5. Figure 9.9 illustrates two leaves of lightweight concrete blockwork plastered or dry-lined (12.5 mm) on each room face. The width of the cavity should be at least 75 mm and the weight of the wall should be at least 250 kg/m^2 (including the plaster or dry lining).

The face of the blockwork should be sealed with cement paint through the full width and depth of any intermediate floor.

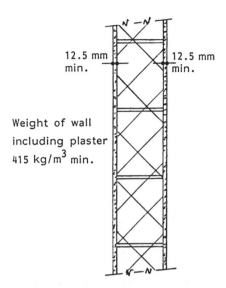

Figure 9.7   Sound insulation – concrete blockwork plastered on both faces

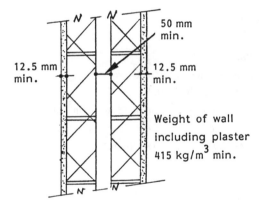

Figure 9.8   Two leaves of brickwork or dense blockwork plastered on external faces

Cavity walls (double cavity)
Example 6. Figure 9.10 illustrates a double cavity wall with a masonry core and lightweight panels either side having a minimum clearance between the masonry core and the free-standing panels of at least 5 mm.

This type of construction should only be used in conjunction with a concrete ground floor to prevent sound escaping via air paths. The core materials can consist of the following materials and minimum weights:

(a)   brickwork – 300 kg/m^2
(b)   concrete blockwork – 300 kg/m^2 (block density 1500 kg/m^3 min)

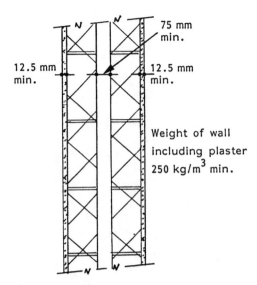

Figure 9.9   Two leaves of lightweight blockwork plastered or dry-lined (12.5 mm min) on each external face

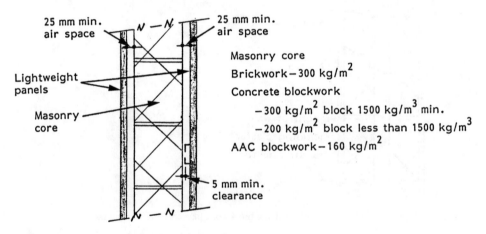

Figure 9.10   Masonry with lightweight panels

(c)   concrete blockwork – 200 kg/m² (block density less than 1500 kg/m³)

(d)   autoclaved aerated concrete blockwork – 160 kg/m².

Sound absorption

Sound absorbents usually only benefit the occupants of the room in which they are used, e.g. a sound absorbent ceiling in an office or factory will reduce the noise level in the room by stopping the sound being reflected. Similarly, furnished rooms used in sound insulation field tests usually give better results than empty rooms.

Refuse chutes

Refuse chutes separating a habitable room can be incredibly noisy and should therefore have a weight (including any plaster finishes) of at least 1320 kg/m^2.

Workmanship and constructional details

Airtightness is a very important factor in the sound insulation of masonry walls and a high standard of workmanship is therefore required.

The minimum thickness of a party wall should always be maintained, particularly where chases, recesses, chimney flues and electrical sockets, etc. are to be built into the wall. Where joists span perpendicular to the party wall they should be supported on joist hangers and not built into the wall. Hollow cored concrete floor units supported on party walls should have their voids filled at the bearings.

Connections between leaves of party walls should be kept to a minimum consistent with structural stability. If butterfly wall ties are not permissible, it is better to use a single-leaf wall.

References

BBA, BRE, BEC and NHBC *Cavity insulation of masonry walls – dampness risks and how to minimise them.*

BS 6232 *Thermal insulation of cavity walls by filling with blown man-made mineral fibre.* Part 1: 1982 *Specification for the performance of insulation systems.* Part 2: 1982 *Code of practice for installation of blown man-made mineral fibre in cavity walls with masonry and/or concrete leaves.* Buildings Research Establishment.

BS 6676: 1986 *Thermal insulation of cavity walls man-made mineral fibre batts (slabs).* Part 1 *Specification for man-made mineral fibre batts (slabs).*

BS 8208: Part 1: 1985 *Assessment of suitability of external cavity walls for filling with thermal insulants – existing traditional cavity construction.* BSI, London.

BS 6213: 1982 *Guide to the selection of constructional sealants.* BSI, London.

BS 8104: 1992 *Methods for assessing exposure of walls to wind driven rain.* BSI, London.

BS 2750: Parts 1 and 4: 1980 *Methods of measurement of sound insulation in buildings and of building elements.* BSI, London.

Buildings Research Establishment Digest 75: 1976 *Cracking in buildings.* Buildings Research Establishment.

Buildings Research Establishment Digest 127 *An index of exposure to driving rain.* Buildings Research Establishment.

Buildings Research Establishment Digest 224 *Cellular plastics for building.* Buildings Research Establishment.

Buildings Research Establishment Digest 236 *Cavity insulation.* Buildings Research Establishment.

Buildings Research Establishment Digest 252 August (1981) *Sound insulation of party walls.* Buildings Research Establishment.

Buildings Research Establishment Digest 277 *Built-in cavity wall insulation for housing.* Buildings Research Establishment.

C & CA Design Guide (1982) *Thermal insulation of masonry walling.*

CIBS Guide A3: 1980 *Thermal properties of building structures.*

The Building Regulations (1985) Approved document E. *Airborne and impact sound.*

Empirical design of free-standing, laterally loaded and internal walls and partitions

10.1   Free-standing masonry walls

Free-standing walls, whether they be in a position of extreme exposure or in a sheltered area, must be designed to withstand the effects of the elements. Unfortunately this type of construction usually receives the minimum of thought at the design stage, yet is open to more abuse generally than the more sophisticated elements of construction which may be adequately protected.

Exposure to wind, rain and frost should be assessed when building in an area with which the designer is unfamiliar, particular attention being paid to the combination of the driving rain index and exposure gradings (CP 121: Part 1: 1973; BRE Report, 1976; BS 5628: Part 3: 1985; BS 1084, 1992) as well as the severity of frosts.

It is very important that the correct details are used to produce an aesthetically acceptable construction and also to afford protection of the wall.

Selection of units and mortar

Selection of the correct unit for free-standing masonry walls is perhaps more important than for many other situations as the wall is usually exposed to the weather on both sides. It is particularly important that satisfactory details are used at the top and bottom of the wall, otherwise exposure in these areas may initiate deterioration of the wall. If the wall has a coping any of the combinations in Table 10.1 should be satisfactory. If the wall has a capping (see definition later in text) any of the combinations in Table 10.2 should satisfy.

In free-standing external walls the lateral strength of the masonry is dependent on the tensile bond between the units and the mortar unless a gravity design approach is adopted. It is therefore essential that a mortar providing the necessary tensile bond be used and selection of this mortar depends upon: (i) the type of unit; (ii) the type of wall; and (iii) the degree of exposure and/or soil conditions.

194

TABLE 10.1 Selection or mortar and units – free-standing walls with copings

Units	Mortar designation
Clay bricks (FN or MN)*	(i) or (ii)
Clay bricks FL or ML	(i) (ii) or (iii)
Calcium silicate bricks classes 3 to 7	(iii)
Concrete bricks compressive strength $\geqslant 7\,\text{N/mm}^2$	(iii)
Concrete bricks any	(iii)

* Sulphate-resisting Portland cement is strongly recommended. Where designation (iii) mortar is used for walls with cappings, the use of sulphate-resisting Portland cement is strongly recommended.

TABLE 10.2 Selection of mortar and units – free standing walls with cappings

Units	Mortar designation
Clay bricks FL (FN)*	(i) or (ii)
Calcium silicate bricks classes 3 to 7	(iii)
Concrete bricks compressive strength $\geqslant 20\,\text{N/mm}^2$	(iii)
Concrete blocks (a) density $> 1500\,\text{kg/m}^3$	(ii)
Concrete blocks (b) dense aggregate unit	(ii)
Concrete blocks (c) compressive strength $\geqslant 7\,\text{N/mm}^2$	(ii)
Concrete blocks (d) most AAC blocks	(ii)

* Sulphate-resisting Portland cement is strongly recommended. Where designation (iii) mortar is used for walls with cappings, the use of sulphate-resisting Portland cement is strongly recommended. Some types of autoclaved aerated concrete blocks may not be suitable.

Copings/cappings

The use of cappings as opposed to copings may be satisfactory and aesthetically preferable to some copings. However, from a durability point of view copings are always preferable.

A coping may be defined as a unit or assemblage which sheds rainwater falling on it clear of all exposed faces of the walling it is designed to protect. A capping, whether flush or projecting, does not incorporate a throating or similar device designed to shed water clear of the walling beneath.

The character and general appearance of a wall may be materially altered by the type of coping or capping used and various examples are illustrated in Figures 10.1 to 10.8. The ideal coping projects either side of the masonry wall, has a 40 mm minimum overhang, including a throating of not less than 13 mm width and which is at least 13 mm from the face of the wall. Alternatively, an overhang as illustrated in Figures 10.5 and 10.6 would be satisfactory. During construction care should be taken to ensure that the throating remains clean and is not filled or bridged by mortar. The top surface should also be sloped to shed rainwater to one side, preferably both, and most important of all it should be durable and preferably impervious. Some copings are pre-treated with a silicone solution to make them water-repellent; such copings should be laid in the traditional way but a mastic joint will be necessary as the silicone treatment tends to prevent adequate bond with the conventional cement mortars.

If copings are of a material having different thermal and moisture movement characteristics to the wall they protect, some provision may be needed to accommodate the resulting differential movements. When brick on edge or similar cappings are used it is recommended that suitable galvanized steel, stainless steel or non-

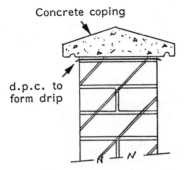

Concrete coping

d.p.c. to
form drip

Figure 10.1   Concrete coping two-way slope

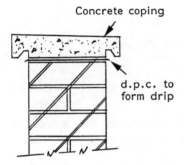

Concrete coping

d.p.c. to
form drip

Figure 10.2   Concrete coping one-way slope

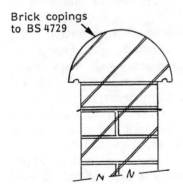

Brick copings
to BS 4729

Figure 10.3   Semi-circular brick coping

ferrous metal end anchors be used to prevent dislodgement of end units (Figure 10.7); if brick on edge cappings are used they must be known to be durable (FL to BS 3921: Part 3: 1985) in such situations and preferably provide an overhang to shed rainwater clear of the wall.

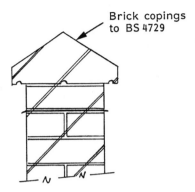

Figure 10.4   Saddle back brick coping

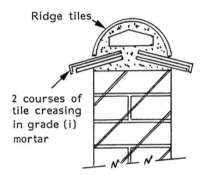

Figure 10.5   Ridge tile coping with creasing tiles

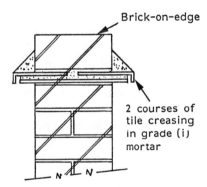

Figure 10.6   Brick-on-edge coping with creasing tiles

Damp proof courses
Whenever possible it is suggested that at the base of a free-standing wall constructed in clay brickwork two or more courses of d.p.c. brick (defined in **BS** 3921: 1985) be used in lieu of other types of d.p.c. membrane which do not provide the same order of resistance to movement. The bricks should be laid in 1:3 cement:sand mortar as

Non-ferrous or galvanized
M.S. fish tail ended cramp

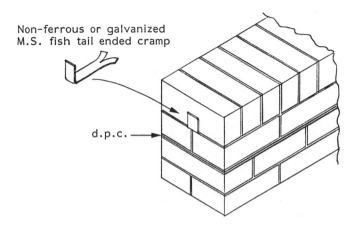

d.p.c.

Figure 10.7   Brick-on-edge coping with end cramp

Brick capping

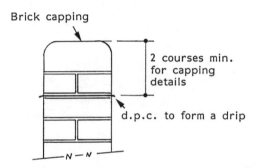

2 courses min.
for capping
details

d.p.c. to form a drip

N — N

Figure 10.8   Brick capping

recommended in BS 743 (1970). Clay d.p.c. bricks should never under any circumstances be bonded with calcium silicate or concrete bricks or blocks.

It should also be noted that masonry d.p.c.s can resist rising damp but will not resist water percolating downwards.

At the base of the wall the d.p.c. should be approximately 150 mm above ground level. If a brick d.p.c. is not appropriate a slate d.p.c. should be the next choice as, like a brick d.p.c., it gives good resistance to movement. The slates should be not less than 230 mm long and have a minimum thickness of 4 mm. A slate d.p.c. is required to consist of at least two courses of slates, laid to break joint, each slate being bedded in 1:3 Portland cement:sand mortar.

If a flexible d.p.c. is used at the base of the wall, allowance must be made for this structural weakness when determining the safe height of the wall.

Rigid d.p.c.s are preferable for two reasons: (a) they provide a certain tensile resistance to flexure; (b) they have a shear resistance of the same order as the wall itself, they provide good resistance to thermal movements and hence, fewer movement joints are required.

At the top of the wall a continuous flexible d.p.c. is recommended in the mortar joint under the coping; if this is omitted rainwater has free access via the mortar joints between the coping units. Flexible d.p.c.s, wherever they occur, should not be cut back from the face of the brickwork and pointed over to disguise their presence. It is essential that such d.p.c.s should project, if possible, 13 mm to form a drip for rainwater, thus protecting the masonry below. Pointing over d.p.c.s should never be permitted as this allows dampness to bypass the d.p.c. and to some extent render it ineffective.

Foundations
Boundary walls do not generally carry any vertical loads other than their own weight and consequently the foundations do not always receive the necessary attention. Foundations for free-standing walls should be located in undisturbed earth or well-consolidated soil at a depth below the frost line. They may be mass or reinforced concrete depending upon site conditions.

Movement joints
Unless walls are constructed in a flexible mortar which does not contain cement (i.e. a lime:sand mortar) movement joints are necessary if cracking of the masonry is to be avoided. Details of the type and frequency of such joints are discussed in Chapter 4. Under no circumstances should copings or cappings bridge movement joints.

Stability of free-standing walls

All walls should be designed so that they have inherent stability against overturning. This may be achieved in different ways according to the type of wall and the function it has to fulfil. A straight free-standing wall of indefinite length may have adequate stability (subject to provision being made for normal movements) if its thickness is sufficient in relation to its height. Zigzag and serpentine walls are more stable than straight walls due to their geometry. The tensile bond between bricks and mortar is of particular importance in free-standing walls and for a given thickness and height straight free-standing walls constructed in certain types of bricks and blocks are likely to be slightly more stable than some other units manufactured with lightweight material or material known to provide a poor bond strength. (If in doubt consult the manufacturer.) Walls can be divided into a series of panels which are stabilized by buttresses, columns, piers or intersecting walls.

BS 5628: Part 3: 1985 gives the rule-of-thumb height-to-thickness ratios shown in Table 10.3 for free-standing, single-leaf walls without piers subject to the following conditions:

(a)   The walls should be subject to wind loads only.
(b)   The ratios do not apply to parapets and balustrades (see BS 6180: 1982).
(c)   The height is the overall dimension above the level of restraint.
(d)   The masonry units should have a compressive strength not less than 3.5 N/mm^2 and a density not less than 1400 kg/m^3 and should follow the recommendations in BS 5628: Part 3: 1985 on durability.
(e)   The walls should be located in an area with many windbreaks, such as a town, city or well-wooded are, i.e. in protection category 3 described in BS 8104 (1992).

(f)   The walls should have a d.p.c. which is capable of developing the same flexural resistance as the remainder of the wall.
(g)   The walls should be single leaf or fully bonded walls with headers. Under no circumstances should units having different characteristics be used to form a composite wall.

TABLE 10.3 Height-to-thickness ratio for free-standing, single-leaf walls without piers

Wind zone	Height-to-thickness ratio
1	8.5
2	7.5
3	6.5
4	6.0

The wind zones are approximately as follows:

1. SE England.
2. England and Wales, roughly from Newcastle to Plymouth excluding the tip of Cornwall.
3. Northern England, Southern Scotland and part of Northern Ireland, West Wales and the tip of Cornwall.
4. The North of Scotland and the North of Northern Ireland.
For more precise locational information, see Figure 1, *Wind Zones* in BS 5628: Part 3: 1985.

Free-standing walls without piers which may sustain greater wind pressures than those described above and/or support other loadings should be designed following the recommendations of BS 5628: Parts 1 (1978) or 2 (1978).

The BDA Booklet 'Design of free-standing walls' (Korff, 1984) gives further information on calculation methods for the above types of wall. Walls with profiled plan formations as illustrated in Figures 10.9 and 10.10 have much greater stability than straight walls and advice on permissible dimensions is also given in the BDA booklet (Korff, 1984).

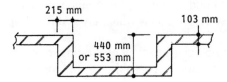

Figure 10.9   Profiled wall plan

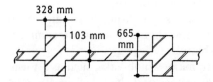

Figure 10.10   Wall plan with piers

Reinforced masonry free-standing walls are outside the scope of this section but further information may be obtained from the BDA booklet (Korff, 1984) or from the BCA (Tovey and Roberts, 1980).

10.2   Walls subjected to lateral loading only

Walls subjected to lateral loading whether loadbearing or non-loadbearing should generally be designed following the recommendations of BS 5628: Parts 1 (1985) or 2 (1978). However, some guidance on the design of non-loadbearing walls subjected to lateral loading is given in BS 5628: Part 3: 1985 and also from numerous other sources (BDA Design Guide, 1986). Before determining the maximum dimensions of a non-loadbearing wall many factors need to be considered, ie:

(a)   the type of unit or if cavity construction units;
(b)   the method of supporting the leaves of the wall, the connections (or lack of connections) may provide a pin-jointed condition, a partially fixed support or a fully fixed support;
(c)   because of differential movement between the masonry and the supporting structure it may be necessary to totally isolate the panel of masonry from the frame-work at the top;
(d)   door and window openings may require the introduction of intermediate supports;
(e)   the shape of the wall may impose dimensional restrictions other than structural;
(f)   the need for provision for thermal/moisture movement may mean that either the panel size has to be restricted or that the introduction of movement joints will affect stability;
(g)   the introduction of reinforcement and/or post-tensioning will have a consider-able affect upon the size of panel required.

Masonry panels used to provide racking resistance for pin-jointed frameworks may come within the accepted definition of 'non-loadbearing' because they are only resisting wind forces. However, such walls need special consideration and should always be designed by a structural engineer.

Table 8 of BS 5628: Part 3: 1985 provides conservative information on maximum areas for certain rectangular walls and gables in buildings up to and including four storeys high.

Example. For Wind Zone 2 for a panel of masonry with three sides fixed and a maximum height of 5.4 m the area of cavity wall permitted is 21 m^2. Similarly for a maximum height of 10.8 m the panel could have an area of 13.5 m^2. The above areas assume the following:

(i)   The building is in an area having a basic wind speed not exceeding 48 m/s and in a zone with numerous windbreaks such as a town, city or well-wooded area (i.e. in protection category 3 described in BS 8104 (1992).
(ii)   The building should not exceed four storeys in height.

(iii)   The walls should be free from any doors, windows or other openings unless intermediate supports are provided or the total area of such openings does not exceed 10 per cent of the permitted area or 25 per cent of the actual area of the wall, whichever is less and no opening is less than half its maximum dimension from the edge of the wall, other than its base, or from any other opening.

(iv)   In a cavity wall:

(a)   the distance between supports should not exceed 30 times the total thickness of the masonry in the wall;

(b)   the thickness of each leaf should be not less than 100 mm excluding plaster or render;

(c)   the cavity width should not exceed 100 mm;

(d)   wall ties should be provided at the rate of not less than 2.5 per m^2;

(e)   the outer leaf should be constructed in bricks or blocks having a compressive strength not less than 14 N/mm^2 and the inner leaf not less than 3.5 N/mm^2

(f)   mortar should not be weaker than designation (iii), i.e. 1:1:6;

(g)   if pitched gable ends have support at the top an equivalent rectangular area may be assumed as shown in Figure 10.11.

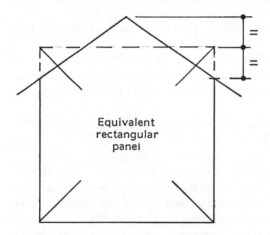

Figure 10.11   Gable walls with edge restraint

Assuming a cavity wall satisfying the above conditions is used to clad a reinforced concrete-framed structure and the standard storey height is 2.7 m, the maximum span would be determined as follows:

Permissible area $= 21 \ m^2$

$\therefore$ span $= \dfrac{21 \ m^2}{2.7 \ m} = 7.78m$

or assuming an inner leaf of 100 mm blockwork and an external leaf of 102.5 mm brickwork

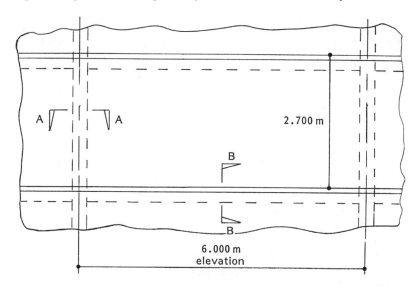

2.700 m

6.000 m
elevation

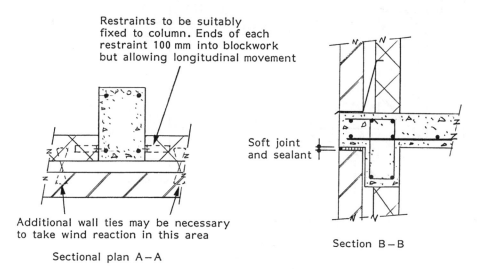

Restraints to be suitably
fixed to column. Ends of each
restraint 100 mm into blockwork
but allowing longitudinal movement

Soft joint
and sealant

Additional wall ties may be necessary
to take wind reaction in this area

Sectional plan A–A

Section B–B

Figure 10.12 Brickwork panel to RC framed construction

$$\therefore \text{span} \qquad = \frac{30(100 + 102.5)}{1000} = 6.08\,\text{m}$$

$\therefore$ The maximum span would be 6 m.

If one of the leaves is increased in thickness to 140 mm the permissible area may be increased by 20 per cent. Assuming 140 mm blockwork and 102.5 mm brickwork the permissible span would be

$$\frac{21\,\text{m}^2 \times 1.2}{2.7\,\text{m}} = 9.34\,\text{m}$$

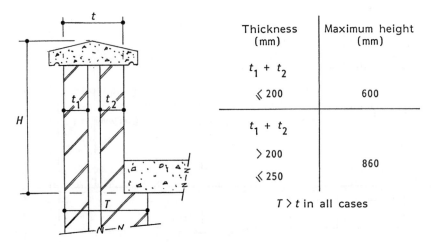

Thickness (mm)	Maximum height (mm)
$t_1 + t_2$ $\ll 200$	600
$t_1 + t_2$ > 200 $\ll 250$	860

$T > t$ in all cases

Figure 10.13 Cavity wall parapet dimensions

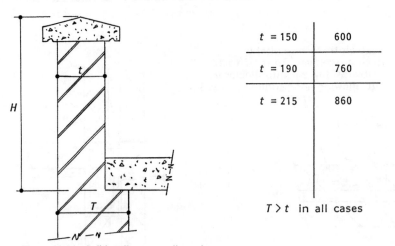

$t = 150$	600
$t = 190$	760
$t = 215$	860

$T > t$ in all cases

Figure 10.14 Solid wall parapet dimensions

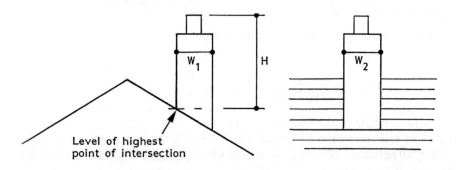

Level of highest point of intersection

$H = 4.5\,W_1$ or W_2 (whichever is the lesser)

Figure 10.15   Chimney heights

or $$\frac{30(140 + 102.5)}{1000} = 7.28 \text{ m}$$

$\therefore$ The maximum span $= 7.25 \text{ m}$ (say)

It is generally more economic to design such walls using engineering principles based on the recommendations of BS 5628: Part 1 (1978) or 2 (1978).

The edge conditions assumed in the above example are illustrated in Figure 10.12.

Parapet walls

Parapet walls should generally be designed and constructed in accordance with the recommendations of BS 5628: Parts 1 (1978) and 3 (1985) using the design criteria from BS 6180 (1982).

The rule-of-thumb height-to-thickness ratios shown in Figures 10.13 and 10.14 can be used for residential buildings (in England) of not more than three storeys and other small buildings of a similar height. Most parapet walls are likely to be severely exposed, irrespective of the climatic exposure of the building as a whole and copings and d.p.c.s should be provided. Notes on suitable masonry units and mortar designations are included in Chapter 3.

Domestic chimneys

In England where a chimney is not adequately supported by ties or restrained in any other appropriate manner, its height measured from the highest point of intersection with the roof (as illustrated in Figure 10.15) should not exceed 4.5 × the least horizontal dimension.

10.3   Internal walls and partitions

All internal walls in buildings with large openings should be designed to resist the appropriate wind forces using the recommendations of BS 5628: Parts 1 (1978) or Part 2 (1985).

In buildings with normal openings and internal non-loadbearing walls the rule-of-thumb proportions quoted in BS 5628: Part 3: 1985 can be used subject to certain conditions:

(a)   the walls must be adequately restrained
    (i)   at both ends, with or without restraint at the top
    (ii)   at the top only.
(b)   consideration should be given to the following factors which may affect stability:
    (i)   accommodation for movement;
    (ii)   openings;
    (iii)   chasing;
    (iv)   exceptional lateral loading if appropriate;
    (v)   wind loading.

If an internal wall is restrained at the top but not at the ends the thickness of the wall may be based on the following empirical formula:

$t \geqslant H/30$

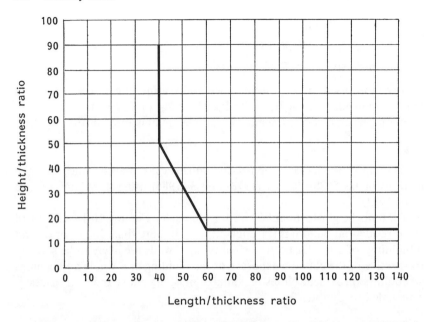

Figure 10.16 Wall panel sizes for restraint at both ends but not the top

The graph in Figure 10.16 for walls restrained at both ends but not at the top is derived from the formulae:

$t \geq L/40$ and $t \geq H/90$ or $t \geq H/15$ with no restriction on the value of L or $t <$ $L/40$ $t > $ L/59 and $t \geq (H + 2L)/133$.

The graph in Figure 10.17 for walls restrained at both ends and at the top is derived from the formulae:

$t \geq L/50$ and $t \geq H/90$ or $t \geq H/30$ with no restriction on the value of L or $t <$ $L/40$ and $t \geq L/110$ and $t \geq (3H + L)/200$. where t is the thickness (mm), H is the height (mm), and L is the length (mm).

If it is known that an internal wall or partition is to be plastered, a maximum thickness of 13 mm of plaster to one side or both sides of the wall may be included when determining the thickness of the wall. Temporary bracing may be necessary for these walls before plastering.

The above formulae assume that the walls are constructed of single-leaf masonry or fully bonded walls with headers. Under no circumstances should units having different characteristics be used to form a composite wall, nor should dissimilar units be used as make-up pieces.

Tables 10.4–10.6 have been calculated on the basis of the formulae suggested in Clause 18.5 of BS 5628: Part 3: 1985 for non-loadbearing walls plastered both sides. For unplastered walls the thicknesses shown should be reduced by 26 mm and rounded off (upwards) to the nearest standard unit thickness.

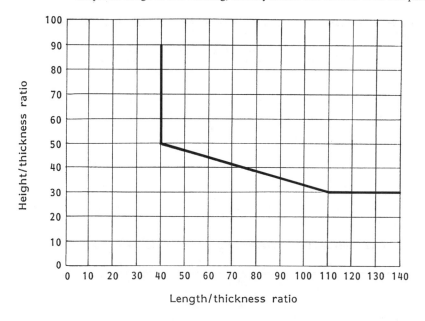

Figure 10.17   Wall panel sizes for restraint at both ends and at the top

TABLE 10.4 Minimum unit thickness (mm) for walls restrained at both ends but not at the top. Walls plastered both sides*

| Height (m) | Length in metres | | | | | | | | |
	4	5	6	7	8	9	10	11	12
2.4	60	75	90	100	140	140	140	140	140
2.7	60	75	90	100	140	140	150	190	190
3.0	60	75	90	140	140	140	150	190	190
3.3	60	75	90	140	140	140	150	190	190
4.0	75	90	100	140	140	140	190	190	190
6.0	+	100	140	140	140	190	190	190	200
8.0	+	+	140	140	190	190	190	200	215
10.0	+	+	+	+	190	190	200	215	230

* Walls plastered both sides, therefore effective thickness equals unit size plus 26 mm.
Some unit manufacturers do not recommend the minimum thicknesses for these height/length ratios.

Provision for services and fittings

Designers should ensure that the stability of walls is not impaired by fixings, chases or holes. This is particularly important when walls are constructed of hollow, cellular or perforated units. In walls constructed of solid units, BS 5628: Part 3: 1985 restricts the depth of horizontal chases normally to one-sixth of the thickness of the single leaf at any point and the depth of vertical chases to one-third of the thickness at any point. Diagonal chases should never be tolerated.

Cutting of holes to accommodate items of equipment should generally be restricted to approximately 300 mm square unless the wall is specifically reinforced in an appropriate manner.

TABLE 10.5 Minimum unit thickness (mm) for walls restrained at both ends and at the top. Walls plastered both sides*

Height (m)	Length in metres								
	4	5	6	7	8	9	10	11	12
2.4	50	50	50	50	50	60	60	60	60
2.7	50	50	50	50	60	60	75	75	75
3.0	50	50	50	60	60	75	75	75	75
3.3	50	50	60	60	75	75	75	90	90
4.0	60	60	75	75	75	90	90	90	100
6.0	+	90	100	100	140	140	140	140	140
8.0	+	+	140	140	140	140	150	150	190
10.0	+	+	+	+	190	190	190	190	190

* Walls plastered both sides, therefore effective thickness equals unit size plus 26 mm.
Some unit manufacturers do not recommend the minimum thicknesses for these height/length ratios.

TABLE 10.6 Minimum unit thickness for walls restrained at top only. Walls plastered both sides

Height (m)	Unit thickness (mm)
2.28	50
2.58	60
3.03	75
3.48	90
3.78	100
4.98	140
5.28	150
6.48	190
6.78	200
7.29	215

* Walls plastered both sides, therefore effective thickness equals unit size plus 26 mm.
Some unit manufacturers do not recommend the minimum thicknesses for these height/length ratios.

Where heavy fittings (such as storage heaters, etc.) are to be fixed to a wall the effect on the stability of the masonry should be considered.

References

BDA Design Guide (1986) *Brickwork to steel framed buildings*. BDA.
BS 5628: Part 3: 1985 *Use of masonry – materials and components, design and workmanship*. BSI, London.
BS 8104: 1992 *Methods for assessing exposure of walls to wind-driven rain*. BSI, London.
BS 3921: 1985 *Clay bricks*. BSI, London.
BS 743: 1970 *Materials for damp proof courses*. BSI, London.
BS 6180: 1982 *Code of practice for protective barriers in and about buildings*. BSI, London.
BS 5628: Part 1: 1978 *Structural use of masonry – unreinforced masonry*. BSI, London.
BS 5628: Part 2: 1985 *Structural use of masonry – structural use of reinforced and pre-stressed masonry*. BSI, London.
Building Regulations (1985) Approved Document A. *Structure*. HMSO.
Buildings Research Establishment Report (1976). *Driving rain index*.
CP 121: Part 1: 1973 *Walling – brick and block masonry*. (Now withdrawn). BSI, London
Kenchington Little Publications. *Safe load tables for laterally loaded masonry panels*.
Korff, J. O. A. (1984) *Design of free-standing walls*. BDA.
Tovey, A. K. and Roberts, J. J. (1980) *Interim design guide for reinforced concrete blockwork subject to lateral loading only*. C & CA Interim Technical Note 6.

11

Calculated loadbearing masonry

The term 'loadbearing masonry' in effect means masonry which is load and self-supporting; BS 5628: Part 1: 1978 interprets it as walls primarily designed to carry an imposed vertical load in addition to their own weight.

The prime requisite of loadbearing masonry is that the floor plan should be repetitive and walls carrying floor and other loads should be taken down to the foundations. Two basic methods of layout can be used, crosswall construction and cellular construction, but in general most designs are a combination of both.

11.1 Methods of construction

Crosswall construction

Crosswall construction (Figure 11.1) is very widely used for low-rise building and is particularly suitable for terraced housing and medium-rise blocks of flats and maisonettes, etc. The crosswall principle is very popular with architects, no doubt because it offers maximum freedom in elevational treatment due to floor and roof loads being carried on party and where appropriate, partition walls.

Crosswalls offer considerable rigidity in their length and usually provide adequate resistance to wind forces in that direction. However, wind forces in the other direction may need to be catered for by return walls, corridor walls, concrete upstands, staircases or liftshafts, etc., if internal piers or buttresses are not present. Curtain walling is often used but it must be emphasized that, unless it is specifically designed to carry wind loading, this form of infilling is no substitute for a return wall, nor is it intended to be.

The use of the crosswall method provides optimum flexibility for planning, rationalization of site operations and the economic usage of masonry units.

Cellular construction

Cellular construction (Figure 11.2) is usually extremely rigid and follows the more traditional pattern of building but when masonry units are used compositely, i.e. in

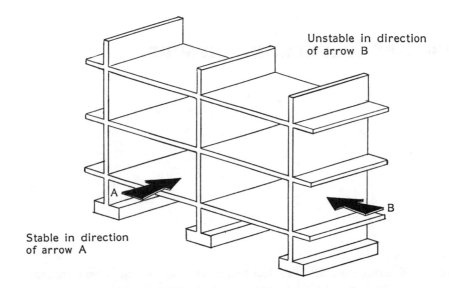

Unstable in direction
of arrow B

Stable in direction
of arrow A

Figure 11.1   Crosswall structure

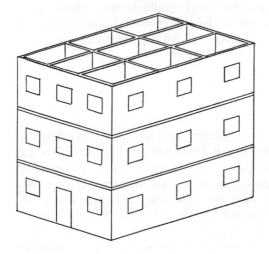

Figure 11.2   Cellular construction

conjunction with other materials (sometimes having limited loadbearing capacity) careful design can produce economic structures making optimum use of the masonry strength. Although the general requirement is for a building with a repetitive plan arrangement, the first one or two storeys may be of the reinforced concrete table form of construction to provide for entrances and other special amenities at foyer level.

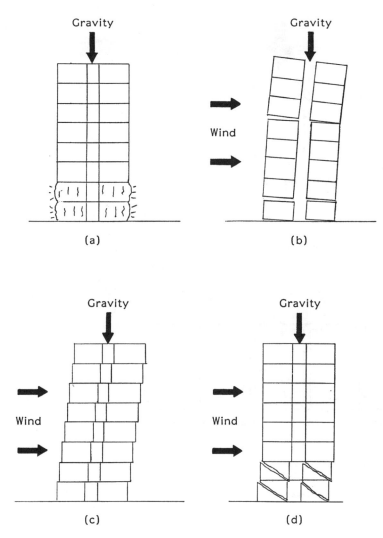

Figure 11.3 Possible conditions of overstressing

The building

In the normal multi-storey loadbearing masonry building four possible conditions of overstressing should be considered as shown in Figure 11.3. The critical conditions in design are usually as shown in Figure 11.3a and 11.3b. Figure 11.3a illustrates excessive vertical compression; Figure 11.3b illustrates the possibility of tension and/or excessive compression developing due to wind loading; Figure 11.3c illustrates sliding of walls and floors at each floor level. This is usually negligible; and Figure 11.3d illustrates a shear or diagonal tension failure due to combined vertical and horizontal loading.

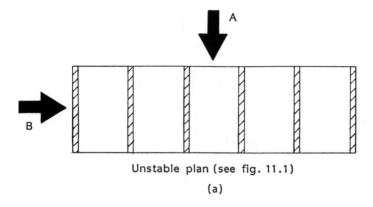

Unstable plan (see fig. 11.1)

(a)

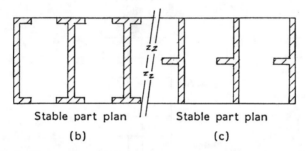

Stable part plan   Stable part plan

(b)   (c)

Figure 11.4   Stable and unstable plan forms

Similarly, Figure 11.4a illustrates an unstable plan whereas Figures 11.4b and 11.4c show alternative stable plans.

Accidental damage

For buildings with five or more storeys the Building Regulations require designers to ensure that the building will not collapse catastrophically as the result of misuse or an accident, i.e. the sort of collapse that occurred at Ronan Point in 1968.

BS 5628: Part 1: 1978 points out that no structure can be expected to be resistant to the excessive loads or forces that could arise due to an extreme cause, but requires that it should not be damaged to an extent disproportionate to the original cause. The items to be considered are therefore:

(a)   robustness and containment of damage;
(b)   resistance to a prescribed horizontal force;
(c)   prevention of vehicular impact;
(d)   special hazards related to service use.

Methods of designing for the above are detailed in a BDA publication (Morton, 1985).

Robustness and containment of damage
To satisfy this condition BS 5628: Part 1: 1978 prescribes specific rules and outlines three options.

Option 1. This requires one structural member at a time to be 'notionally removed' and the structure analysed to predict the extent of collapse. If the volume lost exceeds that permitted by the Building Regulations, the element in question must be strengthened to become what is described as a 'protected member' or the structure improved to reduce the extent of collapse to an acceptable level. A 'protected member' is an element designed to carry without failure a reduced design load and an accidental force equivalent to 34 kN/m^2.

Option 2. This requires the addition of horizontal structural ties and a check on vertical members to ensure that they are 'protected' or, if removed, will not lead to a catastrophic collapse.

Option 3. This requires the addition of horizontal and vertical structural ties.
  BS 5628: Part 1: 1978 also gives rules on containment and members are defined as beams, columns, slabs or walls. If a wall is not a 'protected member' the Standard gives guidance on what amount of wall should be deemed to be removed by suggesting three specific cases where the spread of damage is limited by the provision of vertical bracing in the form of intersecting and return walls; intersecting substantial partitions or piered walls.

Resistance to a prescribed horizontal force
The horizontal tying force, F_t is the lesser of 60 kN or 20 kN + 4 N_s where N_s is the number of storeys. This means that the maximum value for F_t is 60 kN and the minimum value for a five-storey structure is 40 kN.

Prevention of vehicular impact
Where vehicular impact is likely to be a problem the Standard suggests that attempts should be made to isolate the structure from vehicles by using earth banks, retaining walls, crash barriers or bollards.

Special hazards related to service use
The Standard suggests that special precautions should be taken where there are particular hazards such as chemical plants or flour mills.

11.2   Overall stability and minimum stiffness

Masonry buildings have collapsed due to lack of co-ordination of the elements, e.g. one engineer designs the masonry and another the roof. In order to avoid this situation recurring BS 5628: Part 1: 1978 requires that: 'The designer responsible for the overall stability of the structure should ensure the compatibility of the design and details of parts and components. There should be no doubt of this responsibility for overall stability when some or all of the design and details are not made by the same designer'.

BS 5628: Part 1: 1978 assumes that all lateral forces acting on the whole structure are resisted by walls in planes parallel to these forces, or by suitable bracing. Additionally, it requires buildings to be designed to resist a uniformly distributed horizontal load equal to 1.5 per cent of the total characteristic dead load above any level or the wind force whichever is greater.

Structural behaviour of masonry

Masonry walls subjected to axial compression and tested to failure in the laboratory normally fail by vertical splitting as a result of horizontal tension in the masonry, although modes of failure do vary dependent upon the type of unit and mortar used. The reason for this type of failure may not be apparent unless it is appreciated that the mortar and masonry unit usually have widely differing strain characteristics.

It has been suggested that the mortar between high-strength units behaves rather like a pad of rubber and that the rubber, having a high lateral strain when compressed, forces the bricks apart; hence the tensile failure occurs by vertical splitting. The tensile strength of the unit is therefore important.

The above analogy tends to simplify the mechanism of failure, which is influenced by other factors including Poisson's ratio, Young's modulus of elasticity, bond between units and mortar, coefficient of friction between mortar and unit, shear resistance and tensile strength.

Shear failure of walls under axial compression has also been observed in several wall tests. When masonry is loaded eccentrically, failure will normally take place by crushing and spalling on the compression face. This may be relieved to some extent by raking back the joints although this in turn reduces the effective wall area. When weak units and mortar are used and the slenderness ratio and eccentricity is high, failure may occur by buckling.

Basic strength of masonry

The general influence of unit and mortar strength on the compressive strength of masonry has been studied extensively and tests show that the compressive strength increases as the strength of both unit and mortar increases but in neither case in direct proportion.

Unit strength/wall strength
The relationship between unit compressive strength and wall strength might be simple if all units had the same geometry and were tested in the same manner and all walls were of the same height. However, the variables in types of material, methods of manufacture, type of unit (i.e. solid, frogged, perforated. cellular, hollow, etc.) shape factor (height to thickness ratio), crushing strength, etc. all affect the appropriate ratio of unit strength to wall strength, regardless of the mortar designation.

When attempts are made to measure the apparent strength of masonry units, differing values are obtained depending on the shape of the unit. A squat brick-shaped unit will have a much higher apparent strength than a more slender block-shaped unit if manufactured from the same material. The difference is due primarily to the restraint imposed by the loading platens of the testing machine, but other factors such as method of manufacture and properties of the material have an influence. In theory one might expect the resulting masonry to be independent of the shape of the units for the same basic material. However, in practice, taller units for a given width reduce the number of mortar joints in a given height so that the overall effect of unit shape on masonry strength is more complex. The Standard makes provision for the above by providing tables for characteristic compressive strength of masonry based on the compressive strength of the units and four mortar designations. The tables are for standard brick format and for formats of 0.6 and 2.0 to 4.0 height to thickness ratios.

Walls or columns of small plan area
In small areas of masonry there is an increased possibility that sufficient lower than average strength units may be present for the strength of a wall to be affected adversely. BS 5628: Part 1: 1978 therefore recommends the application of a reduction factor to the characteristic compressive strength of walls having an area less than 0.2 m^2.

Narrow brick walls. Research has shown that walls constructed of narrow half bricks (width equal to or greater than the length) have greater resistance to vertical splitting due to axial loading than bonded walls which appear to be weakened by the presence of more vertical mortar joints. The Standard allows an enhancement factor of 1.15 to such walls with the exception of walls constructed with modular bricks. The enhancement factor also applies to the loaded inner leaf of a cavity wall if it is of half brick thickness. Even higher strengths are anticipated for the slender modular bricks and subject to certain conditions the enhancement factor is increased to 1.25.

Walls constructed of wide bricks. When walls are constructed of bricks having a ratio of height to width of less than 0.6 the characteristic compressive strength is required to be determined from wall tests.

Hollow block walls. It is important to lay the units in a manner which ensures that the webs are aligned vertically.

Natural stone masonry. Natural stone masonry is designed on the same basis as solid concrete block of equivalent compressive strength. If the masonry consists of large, carefully shaped pieces with relatively thin joints, design stresses in excess of those quoted for solid concrete may be allowed but the Standard does not give any guidance on enhancement factors.

Random rubble masonry. The strength of random rubble masonry is normally taken at 75 per cent of the corresponding strength of natural stone masonry when constructed in mortar containing Portland cement and 50 per cent of that for masonry constructed in mortar designation (iv) (1:2:8 to 9) or if constructed in a lime mortar (i.e. one not containing Portland cement).

Structural units laid other than on the normal bed face
When structural units are laid other than on the normal bed face the compressive strength of the unit is determined in that aspect in accordance with the appropriate Standard.

Hollow blocks. When hollow blocks are bedded on mortar strips along only the outer edges, load is transmitted only through these strips. The strength of the wall is affected and the permissible strength determined from the Standard should therefore be reduced by the ratio of the bedded area to the net plan area of the block, taking into account the voids, unless the block has been tested on two strips of mortar.

11.3   Design factors

Characteristic compressive strength of masonry

The characteristic compressive strength of masonry is the value of the strength of masonry below which the probability of test results falling is not more than 5 per cent. Figure 1 and Tables 2 (a) (b) (c) and (d) of the Standard (BS 5628: Part 1: 1978) give specific values for characteristic compressive strength of masonry.

Characteristic flexural strength of masonry

The characteristic flexural strength of masonry is based on the same probability of test results as characteristic compressive strength and is only intended for masonry design in bending. However, the Standard does allow up to 50 per cent of the values quoted in direct tension at the designer's discretion when suction forces due to wind loading are transmitted to masonry walls, or when the probable effects of misuse or accidental damage are considered. Flexural tension should not generally be relied upon at d.p.c.s unless test evidence is available to justify such assumptions.

Characteristic shear strength of masonry

The characteristic shear strength quoted in the Standard relates to horizontal shear and makes allowance for dead and imposed loads. One formula is given for masonry constructed in mortar designations (i), (ii) and (iii) and another giving reduced values for mortar designation (iv).

Coefficient of friction

The coefficient of friction between concrete and masonry is quoted in the Standard as 0.6.

Partial safety factors for material strength

$$\text{Design strength} = \frac{\text{characteristic strength } (f_k)}{\text{partial safety factor } (\gamma_f)}$$

The value of the partial safety factor for materials is dependent upon the degree of quality control exercised both during the manufacture of the units and in the site supervision, as well as on the quality of the mortar used during construction. Two levels of control are recognized in each case.

Manufacturing control
Special category may be assumed if the manufacturer is able to supply structural units to a specified compressive strength limit (acceptance limit) which has a probability of not more than 2.5 per cent of being below that limit and if the manufacturer operates an appropriate quality control scheme.

Normal category assumes that the supplier can meet the requirements for compressive strength in the appropriate British Standard but does not meet the other requirements for special category.

Construction control
Special category may be assumed when the designer, either by frequent visits to the site or by the presence of a permanent representative on site, ensures that the work is built in accordance with the appropriate clauses of BS 5628: Part 3: 1985. Also that preliminary compressive strength tests are carried out on the mortar and that the results of such tests comply with the strength requirements of Table 1 in BS 5628: Part 1: 1978.

Normal category is assumed whenever the requirements of special category are not wholly satisfied.

Table 11.1 shows the partial safety factors for the various combinations of manufacturing and site control

TABLE 11.1 Partial safety factors for material strength γ_m

		Category of construction control	
		Special	*Normal*
Category of manufacturing	Special	2.5	3.1
control of structural units	Normal	2.8	3.5

Based on Table 4 of BS 5628: Part 1: 1978

Wall tests
The characteristic compressive strength for masonry can be determined from the ultimate strength of brickwork or blockwork panels tested to destruction. Partial safety factors for such panels are normally taken as 0.9 times the values quoted in Table 11.1.

Design considerations

Slenderness ratio. This is the ratio of the effective height or length to the effective thickness. The ratio should not exceed 21, except for walls less than 90 mm thick, in buildings of more than two storeys, where the maximum is 20.

Effective height or length. The effective height or length of a wall is dependent upon the relative stiffness of the elements of structure connected to the wall or column and

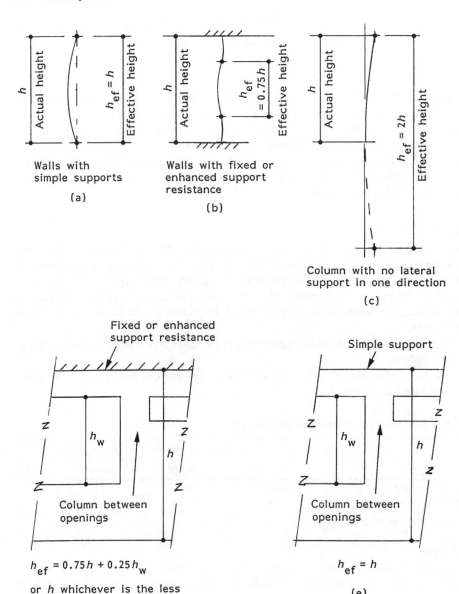

Figure 11.5 Effective height of walls and columns

the efficiency of the connections (see Figures 11.5–11.7). When walls are stiffened by intersecting walls a stiffening coefficient may be used on the assumption that the intersecting walls are equivalent to piers (see Table 5 of the Standard).

Piers. These may be treated as walls for effective height purposes when their thickness is not greater than 1.5 times the thickness of the wall (see Figure 11.8).

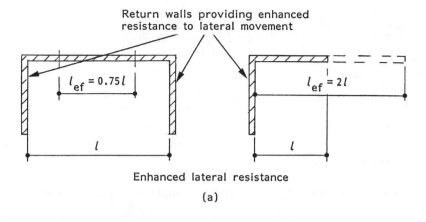

Figure 11.6 Effective length of walls

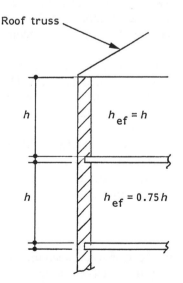

Figure 11.7 Typical example of effective heights

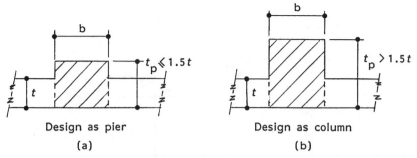

Figure 11.8   Width relating to piers and columns

Effective thickness. The effective thickness of walls, columns or piers is shown in Figure 11.9.

Special items for consideration. As loadbearing masonry is frequently made up of separate leaves of different materials having dissimilar and usually opposing movement characteristics, it is essential that provision for movement be considered (see Chapter 4). When loadbearing walls abut with non-loadbearing walls care should be taken to prevent the accidental transfer of load to the non-loadbearing walls.

11.4   Cavity walls

When load is carried by one leaf of a cavity wall only, the loadbearing capacity of the wall should be based on one leaf only. However, the stiffening effect of the other leaf can be taken into account when calculating the slenderness ratio of the wall. The minimum thickness of each leaf of a cavity wall should be not less than 75 mm.

The width of cavity between the leaves may vary between 50 mm and 150 mm but should not exceed 75 mm where either of the leaves is less than 90 mm. BS 5628: Part 1: 1978 allows a smaller cavity with appropriate supervision but this is not recommended by the writer.

Leaves of cavity walls should be united by metal wall ties and where the width of the cavity exceeds 75 mm only vertical twist-type wall ties should be used. The number of wall ties required per square metre ranges from 2.5 to 4.9, dependent upon the minimum leaf thickness and cavity width (see Table 6 of BS 5628: Part 1: 1978).

The Standards (BS 5628: Part 1: 1978; BS 1243: 1978; BS 5628: Part 3: 1985) recommend a minimum embedment of wall ties in mortar bed joints of 50 mm in each leaf. However, if tolerances of units, wall ties and building are taken into account this may prove to be insufficient and 70 mm minimum embedment has been suggested elsewhere (de Vekey, 1984).

Support of the outer leaf of external cavity walls

To avoid undue loosening of wall ties due to differential movements between the two leaves, the Standard requires the outer leaf to be supported at intervals of not more than every third storey or every 9 m, whichever is less. However, for buildings not

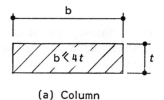

$t_{ef} = t$ or b (depending upon direction of bending)

(a) Column

$t_{ef} = t$

(b) Single leaf wall

$t_{ef} = \frac{2}{3}(t_1 + t_2)$ or t_1 or t_2 (whichever is the greater)

(c) Cavity wall

$t_{ef} = t \times K$ (K = stiffness coefficient)

(d) Single leaf wall stiffened by piers

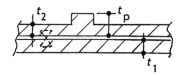

$t_{ef} = \frac{2}{3}(t_1 + Kt_2)$ or t_1 or Kt_2

(e) Cavity wall stiffened by piers

Figure 11.9 Effective thickness of columns and walls

exceeding four storeys or 12 m in height, whichever is less, the outer leaf is permitted to be uninterrupted for its full height. Vertical movement joints are also required for the same purpose (see Chapter 4).

Veneered walls
Walls having a facing which is attached to the backing (but which is not effectively bonded to the backing) are known as veneered walls. The dead weight of such veneer may be taken into account but any structural effect must be neglected.

Faced walls
When a wall is constructed of two different types of masonry unit it should be designed in the same manner as a wall having the same total thickness and built entirely in the weaker material. Faced walls should never be constructed in incompatible materials.

Collar jointed walls
When a solid wall is constructed of two separate leaves with a vertical collar joint not exceeding 25 mm width it may be designed either as a cavity wall or as a single leaf wall subject to conditions described in the Standard (BS 5628: Part 1: 1978).

Walls and columns subjected to vertical loading

The design strength of masonry $= \dfrac{f_k . \beta}{\gamma_m}$

The design vertical load resistance of walls $= \dfrac{\beta t . f_k}{\gamma_m}$

The design vertical load resistance of columns $= \dfrac{\beta b . t . f_k}{\gamma_m}$

where
f_k = the characteristic strength of the masonry (Table 2 of Standard) (BS 5628: Part 1: 1978)

β = the capacity reduction factor which allows for the effects of slenderness and eccentricity (Table 7 of Standard) (BS 5628: Part 1: 1978)

γ_m = the partial safety factor for the material (Table 4 of Standard) (BS 5628: Part 1: 1978)

t = the thickness of a wall or column

b = the width of a column

Effects of eccentricity of loading
It is necessary to determine the effects of eccentricity of loading on walls, whether direct or induced by lateral loading. The methods of determining eccentricity can be

complicated and readers are referred to Appendix B of the Standard also the BDA Handbook on the Standard (Haseltine and Moore, 1981).

Shear forces
Provision against the ultimate limit state in shear being reached may be assumed where

$$V_h \leqslant \frac{f_v}{\gamma_{mv}}$$

where
V_h = the shear stress produced by the horizontal design load calculated as acting uniformly over the horizontal cross-sectional area of the wall
γ_{mv} = the partial safety factor for material strength in shear
f_v = the characteristic shear strength of the masonry

Concentrated loads
Stresses under and close to a bearing.
Guidance is given in the Standard (BS 5628: Part 1: 1978).

Composite action between walls and their supporting beams
Composite action may be taken into account subject to certain conditions:

(a)   the ratio of height to length of the wall must not be less than 0.6 to allow arching to take place;
(b)   openings must not occur within the lines of thrust;
(c)   increased stresses due to composite action must not exceed the local compressive capacity of the masonry.

Example 1
Design a single-leaf brickwork internal wall to carry a concentric load of 145 kN/m run. The height between concrete floors is 2450 mm (Figure 11.10).

Effective height h_{ef} $= 0.75 \times 2450$ mm

        $= 1837.50$ mm

Effective wall thickness $=$ actual thickness

        $= 102.5$ mm

Slenderness ratio $= \dfrac{1837.50 \text{ mm}}{102.5 \text{ mm}}$

        $= 17.93$, say 18

Eccentricity at right angles to the wall $= e_x = 0$.

As the wall is concentrically loaded $\beta = 0.77$ (see Table 7 of the Standard) (BS 5628: Part 1: 1978).

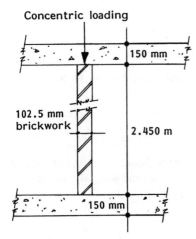

Figure 11.10   Design example 1 – loadbearing walls

Design vertical resistance of wall $= \dfrac{\beta \times t \times f_{k}}{\gamma_{m}}$

The brickwork is assumed to be constructed of normal category bricks under normal category construction control

$\gamma_{m} = 3.5$ (see Table 11.1)
Design vertical resistance of wall

$$= \frac{0.77 \times 102.5 \times f_{k}}{3.5}$$

$$= 22.55\, f_{k}$$

Design vertical load $= 145$ kN/m

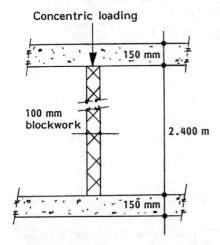

Figure 11.11   Design example 2 – loadbearing walls

$$\therefore f_k = \frac{145 \times 10^3}{22.55 \times 10^3} = 6.43 \text{N/mm}^2$$

From Table 2(a) of the Standard the following minimum combinations of brick and mortar could be used:

(a)   20 N/mm^2 brick in mortar designation (i)
(b)   27.5 N/mm^2 brick in mortar designation (ii) or (iii)
(c)   35.0 N/mm^2 brick in mortar designation (iv)

Note: Mortar designation (iv) is rarely used with bricks for loadbearing brickwork.

Example 2
Design a single-leaf 100 mm blockwork internal wall to carry a concentric load of 45 kN/mm run. The height between concrete floors is 2400 mm (Figure 11.11).

Effective height h_{ef} = 0.75 × 2400 mm
                                                     = 1800 mm
Effective wall thickness   = 100 mm

Slenderness ratio $= \dfrac{1800 \text{ mm}}{100 \text{ mm}} = 18$

Eccentricity at right angles to the wall = $e_x = 0$.

As the wall is concentrically loaded β = 0.77 (see Table 7 of the Standard).

Design vertical resistance of wall $= \dfrac{\beta \times t \times f_k}{\gamma_m}$

Assuming γ_m = 3.5 as for Example 1.

Design vertical resistance of wall $= \dfrac{0.77 \times 100 \times f_k}{3.5} = 22.0 \, f_k$

Design vertical load   = 45 kN/m

f_k required $= \dfrac{45 \times 10^3}{22.0 \times 10^3}$

$$= 2.05 \text{ N/mm}^2$$

Using 215 mm high × 440 mm long solid concrete blocks (ratio of height to least horizontal dimension = 215/100 = 2.15) therefore from Table 2 (d) of the Standard, a block with a unit compressive strength of 2.8 N/mm^2 in any mortar designation may be used.

Example 3
Determine the vertical resistance of the inner leaf of a wall in a multi-storey load-bearing brickwork building for the loading shown in Figure 11.12. Assuming $W_1 = 25$ kN/m, $W_2 = 175$ kN/m and $W_3 = 30$ kN/m.

This example has been simplified to illustrate the method of considering the vertical resistance of a wall in a multi-storey building and the characteristic loads include the effect of wind forces.

Loads W_1 and W_2 are assumed to act at the positions indicated in Figure 11.12 and load W_3 acts at the critical section of the wall for design for compressive strength which is slightly above mid-height between the floors. As the wall is relatively highly stressed both leaves will be constructed in brickwork.

Slenderness ratio. The floors are assumed to be constructed of *in situ* reinforced concrete bearing on the inner leaf of brickwork and consequently provide enhanced lateral resistance to the wall.

$$\text{Effective height } h_{ef} = 0.75 \times 2400 \text{ mm}$$
$$= 1800 \text{ mm}$$

Effective thickness t_{ef} of the wall assuming two leaves of 102.5 mm brickwork

$$= 0.67 (2 \times 102.5) \text{ mm}$$
$$= 137.35 \text{ mm}$$

$$\text{Slenderness ratio} = \frac{h_{ef}}{t_{ef}} = \frac{1800 \text{ mm}}{137.35 \text{ mm}}$$

$$= 13.11$$

Eccentricity of loading. The eccentricity of the floor loading is assumed to be $t/6$ as shown in Figure 11.12 and the combined eccentricity of loads W_1 and W_2 are

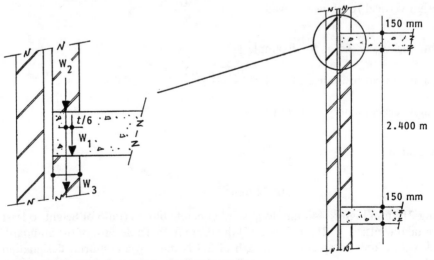

Figure 11.12   Design example 3 – loadbearing walls

determined by taking moments about the centre line of the inner leaf of brickwork hence

$$(175 + 25)\, e_x \quad = \quad 25 \times t/6$$

$$e_x = \frac{25.t}{6 \times 200} \quad = \quad 0.021\; t$$

Table 7 of the Standard quotes capacity reduction factors (β) which allow for the effects of slenderness ratio and eccentricity of loading at the top of the wall.

Hence for a slenderness ratio of 13.11 and an eccentricity of up to

$$0.05t, \quad = \quad \beta\; 0.91$$

Design vertical resistance of wall. As the thickness of the inner leaf of the wall is equal to the width of a standard format brick, the value of f_k may be enhanced by 15 per cent and assuming γ_m to be 3.5 (i.e. normal category bricks and construction) the design vertical resistance of the wall

$$= \frac{\beta t (1.15 f_k)}{\gamma_m} \qquad = \frac{0.91 \times 102.5 \times 1.15 f_k}{3.5}$$

$$= 30.65 f_k$$

The design vertical load $= W_1 + W_2 + W_3 \quad = 175 + 30 \text{ kN/m}$
$$= 230 \text{ kN/m}$$
$\therefore$ Minimum characteristic compressive strength f_k required

$$= \frac{230 \times 10^3}{30.65 \times 10^3} \quad = 7.5 \text{ N/mm}^2$$

From Figure 1 of the Standard, Class 4 bricks in mortar designation (ii) or Class 5 bricks in mortar designation (iii) may be used.

Special control. If it is possible to use both special control of manufacture of the masonry units and special control of construction savings can be made by adopting a partial safety factor for material strength (γ_m) of 2.5

$\therefore$ Design vertical resistance of wall

$$= \frac{0.91 \times 102.5 \times 1.15 f_k}{2.5} \quad = 42.91\, f_k$$

Design vertical load as before $= 230 \text{ kN/m}$

$$f_k \text{ required} = \frac{230 \times 10^3}{42.91 \times 10^3} \quad = 5.36 \text{ N/mm}^2$$

$\therefore$ From Figure 1 of the Standard a Class 3 brick in mortar designation (iii) may be used instead of the Class 4 brick required for $\gamma_m = 3.5$.

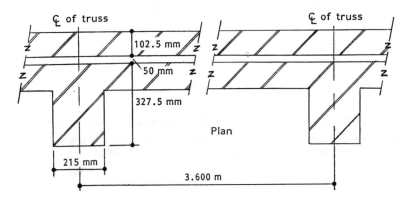

Figure 11.13   Design example 4 – loadbearing walls

Example 4

Design brickwork pier supports for an industrial building consisting of 4.0 m high, 225 mm cavity brickwork external walls and a steel-sheeted roof supported by steel roof trusses at 3.6 m centres.

It is assumed that the roof trusses are fixed to padstones at the top of the piers arranged to provide an axial loading of 130 kN. The plan arrangement of the external wall is illustrated in Figure 11.13.

Pier design. Assume a pier size of 327.5 mm by 215 mm

$$\frac{\text{Thickness of pier}}{\text{Thickness of inner leaf of cavity wall}} = \frac{327.5 \text{ mm}}{102.5 \text{ mm}} = 3.2$$

This is greater than 1.5; therefore the pier should be treated as a column in the plane at right angles to the wall.

Slenderness ratio. The slenderness ratio of the pier will only need to be considered in one plane, as the wall is adequately restrained in the plane of the wall.

Effective height of pier (h_{ef}) = 4000 mm
Effective thickness of pier (t_{ef}) = 327.5 mm

$$\text{Slenderness ratio} \quad = \frac{4000 \text{ mm}}{327.5 \text{ mm}} = 12.21$$

Design vertical load resistance of pier. From Table 7 of the Standard for a slenderness ratio of 12.21 and assumed axial loading the capacity reduction factor $\beta = 0.92$.

Assuming normal categories of manufacturing and construction control the partial safety factor for material strength $\gamma_m = 3.5$

$$\text{Design vertical load resistance} \quad = \frac{\beta.b.t.f_k}{\gamma_m}$$

$$= \frac{0.92 \times 215 \times 327.5 f_k}{3.5} \qquad = 18.5 \, f_k \text{ kN}$$

As the pier is assumed to be correctly bonded to the inner leaf of the cavity wall it will not be necessary to apply the reduction factor for small plan areas of masonry to the characteristic compressive strength f_k.

$$f_k \text{ required } = \frac{130 \times 10^3}{18.5 \times 10^3} = 7.03 \text{ N/mm}^2$$

From Figure 1 of the Standard a Class 3 brick in mortar designation (i) may be used. **Note:** such a design would also need to be checked for lateral loading.

Walls subjected to lateral load

The design of walls to resist lateral forces is based on research data which indicate that well-defined crack patterns occur when walls are tested to failure. Figure 11.14 illustrates the typical crack pattern of a panel wall with fully restrained edges. The Standard (BS 5628: Part 1: 1978) makes it clear that walls subjected mainly to lateral loads are not capable of precise design and suggests that such walls be designed as either:

(a)   panels supported on a number of sides.
(b)   walls acting as an arch between supports.

   Walls which include openings require special treatment.

Design bending moments in panels
As masonry walls are not isotropic there is an orthogonal strength ratio μ (the ratio of the flexural strength of masonry when failure is parallel to the bed joints to that when failure is perpendicular to the bed joints), depending on the brick or block and mortar used. The design bending moment per unit height of a panel in the horizontal direction is taken as: $\alpha.W_k.\gamma_f.L^2$

   where

   α = bending moment coefficient taken from Table 9 of the   Standard (BS 5628: Part 1: 1978)
   γ_f = the partial safety factor for loads
   L = the length of the panel between supports
   W_k = the characteristic wind load per unit area

   When vertical load acts so as to increase the flexural strength this may be taken into account by modifying the orthogonal ratio. For walls spanning vertically, the design moment per unit length of wall at mid-height is taken as:

$$\frac{W_k.\gamma_f.h^2}{8}$$

Calculation of design moment of resistance of panels

The design moment of resistance $= \dfrac{f_{kx}}{\gamma_m}.Z$

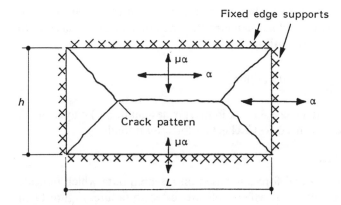

Figure 11.14 Crack pattern – laterally loaded wall

where

f_{kx} = the characteristic flexural strength from Table 3 of the Standard
γ_m = the partial safety factor for materials
Z = the section modulus

Arching
Calculations should be based on a simple three-pin arch and the bearing at the supports and at the central hinge should be assumed as 0.1 times the thickness of the wall.

The maximum design arch thrust per unit width may be assumed to be

$$= \frac{1.5f_k}{\gamma_m} \left(\frac{t}{10}\right)$$

Where the lateral deflection will be small and can be ignored, the design lateral strength is given by:

$$q_{lat} = \frac{f_k}{\gamma_m} \left(\frac{t}{L}\right)^2$$

where

q_{lat} = the design lateral strength per unit area of wall
t = the overall thickness of wall
f_k = the characteristic comprehensive strength of the masonry
L = the length of the wall
γ_m = the partial safety factor for materials

The supporting structure must be capable of resisting the arch thrust with negligible deformation.

Cavity walls

When cavity walls are tied with vertical twist-type wall ties or butterfly/double triangle ties capable of transmitting the compressive forces to which they are subjected the design moment of resistance of the cavity wall may be taken as the sum of the design moment of resistance of the two leaves.

11.5 Support conditions and continuity

When designing masonry panels to resist lateral forces it is necessary to make an assessment of the fixity of the supports and Figures 11.15 and 11.16 give typical examples.

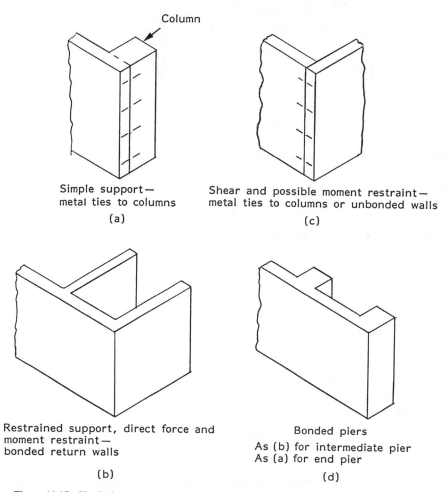

Figure 11.15   Vertical support conditions for laterally loaded panels

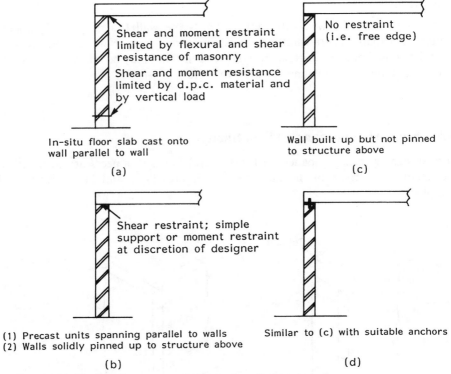

Figure 11.16 Horizontal support conditions for laterally loaded panels

Limiting dimensions

The Standard limits the size of panels as follows:

(a) Panel supported on three sides
    (i) two or more sides continuous:
        height × length equal to 1500 t_{ef} or less
    (ii) all other cases:
        height × length equal to 1350 t_{ef} or less
(b) Panel supported on four edges
    (i) three or more sides continuous:
        height × length equal to 2250 t_{ef} or less
    (ii) all other cases:
        height × length equal to 2025 t_{ef} or less
(c) Panel simply supported at top and bottom. Height equal to 40 t_{ef} or less.
(d) Freestanding wall. Height equal to 12 t_{ef} or less.

In cases (a) and (b) no dimension should exceed 50 times the effective thickness.

Example 5
Design the cladding for a multi-storey steel framed building with dimensions and assumed support conditions as illustrated in Figure 11.17. The panel to be consid-

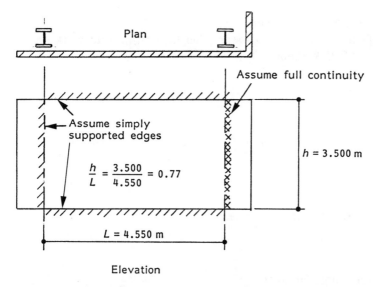

Figure 11.17   Design example 5 – laterally loaded wall

ered is of cavity construction and is required to withstand a lateral wind force of
1.2 kN/m^2. It is assumed that the wall will have an overall thickness of 303 mm and
that the inner leaf will consist of 3.5 N/mm^2 blocks 150 mm thick laid in mortar
designation (iii) and the outer leaf clay bricks 102.5 mm thickness having an absorption less than 7 per cent and laid in mortar designation (iii).

Limiting dimensions. For a panel supported on four sides with less than three sides
continuous, the height × length of the panel must be less than or equal to 2025 t_{ef}

t_{ef} = 0.67 (102.5 + 150) = 169.2 mm
∴   2025 × 169.2^2 = 58 × 10^6 mm^2
$h × L$ = 3500 × 4550   = 15.93 × 10^6 mm^2
Maximum wall dimension must not exceed 50 t_{ef}
∴   50 × 169.2 mm   = 8460 mm
Wall length of 4550 mm is satisfactory.

Characteristic flexural strength. Outer leaf: for brick of less than 7 per cent absorption in mortar designation (iii). From Table 3 of the Standard, Amendment No 3:
1985.

f_{kx} =   0.5 N/mm^2 in the plane of failure parallel to the bed joints
f_{kx} =   1.5 N/mm^2 in the plane of failure perpendicular to the bed
joints.
Inner leaf For 3.5 N/mm^2 block in mortar designation (iii)
f_{kx} =   0.22 N/mm^2 in the plane of failure parallel to the bed joints
(interpolated value for 150 mm thick blocks)
f_{kx} =   0.38 N/mm^2 in the plane of failure perpendicular to the bed joints
(interpolated value for 150 mm thick blocks)

Vertical load due to self weight of outer leaf

= partial safety factor for load $(\gamma_f) \times \frac{1}{2}$ storey height $\times$ wt/m^2 brickwork

$= 0.9 \times 1.75$ mm $\times 2.25$ kN/mm^2

$= 3.54$ kN/m

$\therefore$ Stress due to vertical design load

$$= \frac{3.54 \times 1000}{102.5 \times 1000} = 0.035 \text{ N/mm}^2$$

Modified orthogonal ratio $= \dfrac{f_{kx}(\text{parallel}) + \text{vertical stress} \times \text{panel height}}{f_{kx}(\text{perpendicular})}$

$$= \frac{0.5 + (0.035 \times 3.5)}{1.5} \qquad = 0.42$$

Vertical load due to self weight of inner leaf

$= 0.9 \times 1.75 \times 1.15$ kN/m^2

$= 1.81$ kN/m

$\therefore$ Stress due to vertical design load

$$= \frac{1.81 \times 1000}{150 \times 1000} = 0.012 \text{ N/mm}^2$$

$\therefore$ Modified orthogonal ratio

$$= \frac{0.22 + (0.012 \times 3.5)}{0.38} = 0.69$$

Design moment of resistance of outer leaf

$$= \frac{f_{kx}}{\gamma_m}.Z = \frac{1.5}{3.5} \times \frac{1000 \times 102.5^2}{6} \times \frac{1}{10^6}$$

$= 75$ kN/m/m

Design moment of resistance of inner leaf

$$= \frac{0.38}{3.5} \times \frac{1000 \times 150^2}{6} \times \frac{1}{10^6}$$

$= 0.41$ kN/m/m

$\therefore$ Load taken by outer leaf

$$= \frac{0.75}{0.75 + 0.41} \times 100 = 65 \text{ per cent}$$

$\therefore$ Load taken by inner leaf

$= (1 - 0.65) \times 100 = 35$ per cent

Bending moment coefficients. Outer leaf

$$\frac{h}{L} = \frac{3500}{4550} = 0.77 \mu = 0.42$$

$\therefore$ From Table 9 (condition F) of the Standard the bending moment coefficient

$= 0.042$

Inner leaf $\qquad \dfrac{h}{L} \quad = \quad 0.77\mu \quad = \quad 0.69$

From Table 9 of the Standard the bending moment coefficient
$= 0.032$

Design bending moment $= \alpha . W_k . \gamma_f . L^2$

$= 0.042 \quad \left(\dfrac{0.75}{0.75 + 0.41} \right) \quad 1.2 \times 4.55^2$

$= 0.68 \text{ kN/m/m}$

This is less than the moment of resistance (0.75 kN/m/m)

$\therefore$ Satisfactory

Inner leaf. Design bending moment

$= 0.032 \left(1.2 - \dfrac{0.75}{0.75 + 0.41} \right) 4.55^2$

$= 0.36 \text{ kN/m/m}$

This is less than the moment of resistance (0.41 kN/m/m)

$\therefore$ Satisfactory

Method of design for free-standing walls
Free-standing walls should be designed as cantilevers, except for wall panels between piers or buttresses which may be designed as three-sided or horizontally spanning. If the latter approach is adopted the piers or buttresses must be capable of providing the necessary lateral restraint.

Calculation of design moment in free-standing walls
The design moment of a free-standing wall subjected to horizontal forces

$= W_k . \gamma_f . \dfrac{h^2}{2} + Q_k . \gamma_f . h_L$

where
$\quad W_k =$ the characteristic wind load per unit area
$\quad \gamma_f \;=$ the partial safety factor for loads
$\quad h \;\;=$ the clear height of the wall or pier above restraint
$\quad Q_k =$ characteristic imposed load
$\quad h_L \;=$ the vertical distance between the point of application of the horizontal load Q_k and the lateral restraint.

Calculation of design moment of resistance of free-standing walls
The design moment of resistance across the bed joints

$= \left(\dfrac{f_{kx}}{\gamma_m} + g_d \right) Z$

where

f_{kx} = the characteristic flexural strength at the critical section (which may be the damp proof course)

γ_m = the partial safety factor for materials

Z = the section modulus

g_d = the design vertical dead load per unit area.

When the flexural strength of the masonry cannot be relied upon (i.e. some d.p.c.s may delaminate or provide zero tensile resistance) a free-standing wall can only be used when there is sufficient vertical load acting. The design moment of resistance should then be assessed assuming that the vertical load is resisted by a rectangular stress block when the moment in a wall of uniform thickness

$$= \frac{n_w}{2} \left[t - \frac{n_w \cdot \gamma_m}{f_k} \right]$$

where

t = the thickness of the wall

n_w = the design vertical load per unit length of wall

f_k = the characteristic compressive strength of masonry

Example 6

Determine the minimum thickness of a free-standing wall 2 m high designed to resist a wind pressure of 0.75 kN/m².

It is assumed that the wall will be constructed of clay bricks having a water absorption of less than 7 per cent and that the category of manufacture will be special. A mortar designation (i) $1:\frac{1}{4}:3$ is assumed, as well as normal category of construction control. Therefore γ_{mv} = 2.5. (See Clause 27.4 of the Standard) (BS 5628: Part 1: 1978). Assume density of brickwork = 2000 kg/m³ (1 kgf = 9.81 N).

In view of the above assumptions γ_m = 3.1 from Table 4 of the Standard and the characteristic flexural stress f_{kx} = 0.7 N/mm² (for brickwork subject to a plane of failure parallel to the bed joints) (see Figure 11.18).

To provide the worst combination of loads, γ_f is taken as 1.2 for the wind load and 0.9 for the self weight of the wall.

BS 3921 (1985) permits bricks having an absorption of 7 per cent or less to be used as a damp proof course for external works. It is therefore not necessary to include an additional d.p.c. in this free-standing external wall.

Design for bending. Assuming a wall thickness of t

The section modulus $Z = \dfrac{1000t^2}{6} = \text{mm}^3/\text{m}$

The design bending moment

$$M = W_k \cdot \gamma_f \cdot \frac{h^2}{2}$$

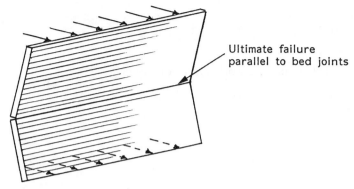

Bending about a horizontal axis

(a)

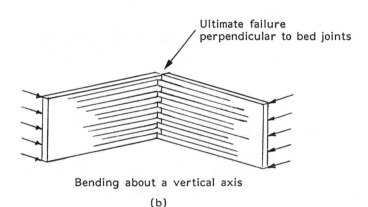

Bending about a vertical axis

(b)

Figure 11.18   Bonding about both axes

$$= 0.75 \times 1.2 \times \frac{2.0^2}{2}$$

$$= 1.8 \text{ kN/m/m}$$

The design moment of resistance

$$= \left(\frac{f_{kx}}{\gamma_m} + g_d \right) Z$$

$$= \left(\frac{0.7}{3.1} + \frac{0.9 \times 2000 \times 9.81 \times 2.0 \times 1}{1000 \times 1000t} \right) \times \frac{1000t^2}{6} \times \frac{1}{10^6}$$

$$= \frac{44.2t^2}{10^6} \text{ kN/m/m}$$

$\therefore$ When the design bending moment = the design moment of resistance

$$1.8 = \frac{44.2t^2}{10^6}$$

$$t = \left[\frac{1.8 \times 10^6}{44.2}\right]^{\frac{1}{2}}$$

t = 202 mm or 215 mm, the work size of one brick.

Design for shear. Design shear force $= \gamma_f . W_k . A_w$

$$= 1.2 \times 0.75 \times 2.0$$

$$= 1.8 \text{ kN/m}$$

Design shear resistance $= f_v . \dfrac{A_w}{\gamma_m}$

$$= \frac{0.35 + \dfrac{(0.6 \times 0.9 \times 2000 \times 9.81 \times 2.0 \times 0.215)}{1000 \times 215} \times \dfrac{215 \times 1000}{1}}{2.5 \times 10^3}$$

$$= 31.9 \text{ kN/m}$$

This exceeds the design shear force therefore a 215 mm wall will be satisfactory.

References

BS 5628: Part 1: 1978 *Structural use of masonry – unreinforced masonry*. BSI, London.
BS 5628: Part 3: 1985 *Structural use of masonry – materials and components design and workmanship*. BSI, London.
BS 1243: 1978 *Metal ties for cavity wall construction*. BSI, London.
BS 3921: 1985 *Clay bricks*.
BS 5390: 1976 *Code of practice for stone masonry*. BSI, London.
BS 6073: Part 1: 1981 *Precast concrete masonry units. Specification for precast concrete masonry units*. BSI, London.
BS 6073: Part 1: 1981 *Precast concrete masonry units. Method of specifying precast concrete masonry units*. BSI, London.
BS 187: 1978 *Calcium silicate (sandlime and flintlime bricks)*. BSI, London.
de Vekey, R.C. (1984) *Performance specifications for wall ties*. Buildings Research Establishment
Haseltine, B. E. and Moore, J. F. A. (1981) BDA Handbook to BS 5628: Part 1. *Structural use of masonry*. BSI, London.
Morton, J. (1985) *Accidental damage robustness and stability*. BDA.

12

Fin and diaphragm walls

12.1 Masonry fin walls

Fin or bonded buttress walls have been used for centuries to resist lateral forces and to provide added stability to loadbearing masonry. However, modern masonry fin walls are now frequently used for tall single-storey structures such as assembly halls, warehouses, theatres, gymnasiums, garages and churches, etc. and offer an economic and often preferred alternative to traditional steel or concrete framed buildings clad in masonry.

The fin wall has been described by Curtin *et al.* (1980, 1984) as an element in structural masonry which has been 'engineered' to exploit masonry's high compressive strength and overcome its low tensile resistance. The fins or deep piers are used to reinforce solid or conventional cavity walls when comparatively large lateral forces are imposed on the walls (e.g. forces due to wind or soil pressure) with little or no applied axial loading.

Types of fin

Fin walls, in effect, consist of a series of 'T'-formed sections of masonry having a flange (the actual wall) and fin (the deep pier) (Figure 12.1). The fin should preferably be fully bonded to the flange (Figure 12.2a) or alternatively connected with sufficient wall ties to resist the vertical shear between the fin and the flange (Figure 12.2b).

When the wall is of cavity construction the two leaves should be tied together in the normal manner. The internal leaf of masonry is not taken into account when determining the moment of resistance of the 'T' section but it does provide some stiffening effect to the assumed flanges. The continuous flange must be capable of providing lateral resistance between the fins if the walls are to perform in a satisfactory manner. Special hollow fins can be used to house services and rainwater downpipes (Figure 12.3) and double fins where provision for movement is required (Figure 12.4).

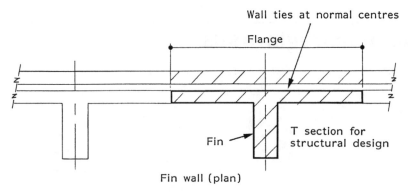

Fin wall (plan)

Figure 12.1   T Section for structural design

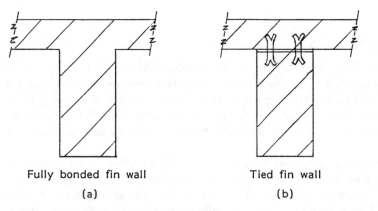

Figure 12.2   Fully bonded and tied fin walls

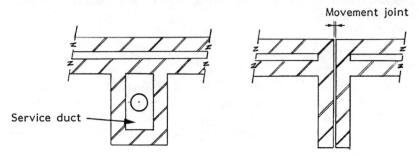

Figure 12.3   Hollow fin

Figure 12.4   Double fin

In elevation the simplest is the rectangular or parallel fin (Figure 12.5a) but tapered fins (Figure 12.5b) and stepped fins (Figure 12.5c) or bevelled fins (Figure 12.5d) are common. More complicated perforated fins can also be used if desired, but these naturally add to the cost of the walls and are more difficult to construct. Port hole fins and perforated tapered fins are shown in Figures 12.5e and 12.5f, respectively.

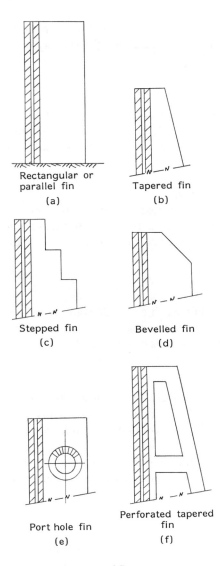

Figure 12.5 Types of fin

Applications

Tall single-storey structures
The current vogue is to use the fin wall technique for this type of structure. The
overall depth of the fin and flange is likely to be greater than that of framed construction, but providing the roof can be used as a horizontal diaphragm or plate, it
can be an economical form of construction with considerable aesthetic appeal.
However, like all forms of structural masonry it demands a high standard of workmanship and quality control.

The fins are normally external but internal fins can be used and if the building requires a crane system occasional internal fins may be necessary for runway beam bearings.

Multi-storey structures

External fin walls have been used for multi-storey loadbearing masonry (particularly in the USA), first, to provide lateral support and second, as an architectural feature.

The 'T' section can also be used for open-plan structures to provide local stiffness, floor-to-floor, against lateral wind forces in addition to providing support to vertical loading. Such systems can be used to advantage in conjunction with internal masonry columns which have limited lateral restraint.

Reinforcement and/or post-tensioning can also be used to produce smaller sections or to resist larger lateral forces. Both techniques have been used in seismic zones and can also be used to advantage to resist accidental forces caused by explosions, etc. (Morton, 1985).

Retaining walls

Buttressed or fin walls are frequently used to retain earth or to add stability to free-standing boundary walls (Korff, 1984). In situations where space is restricted or large lateral forces are imposed, pre-stressed fin retaining walls can be used to advantage.

Remedial work

When existing walls need to be strengthened to increase stability this may be achieved by the addition of fins. Curtin *et al.* (1984) point out that it is not advisable to construct fin walls by block bonding the fins into the continuous wall forming the flanges. If tooth bond is used, special precautions must be taken to ensure that the toothing is fully packed with mortar and the cutting out neatly done. As an alternative, the fins could be fixed to the existing wall with metal shear ties, designed to cater for the vertical shear stress.

Foundations

Generally for single-storey buildings simple strip footings can be used or some form of slab edge-thickening detail as vertical axial loading and overturning moments at the base are relatively small. Even when post-tensioned fins are used, the special details at the base can be accommodated within a modified edge-thickening detail.

Movement joints

Provision for movement must be made in long fin walls as for traditional construction (see Chapter 4). Also, when fin walls are designed as single-storey buildings of large span, account must be taken of the effects of expansion and contraction of the roof structure on the walls. It may be necessary to, for example, provide vertical slip planes or movement joints at the corners of the building. In the length of the building vertical movement joints can easily be accommodated at openings or by the provision of double fins, one at each side of the joint (see Figure 12.4).

Large openings
To prevent problems of high local stresses due to horizontal wind loading and localized axial loads at beam bearings, extra fins or thicker fins can be used each side of openings to support the beams or lintels. Clerestorey lights are frequently used with this type of construction and these can be provided between fins. When walls are post-tensioned clerestorey glazing is also possible and the walls act as free-standing cantilevers.

Thermal insulation
The thermal transmission value (U-value) is the same as that of a normal cavity wall and its value can be improved by the use of cavity insulation and/or thermal blocks internally.

Damp proof courses
Sliding at the base may be the critical feature for low retaining walls which are required to resist high lateral loads. To avoid this a high friction d.p.c. should be used, e.g. for clay-brick fin walls a brick d.p.c. should be used. Some d.p.c.s tend to creep and squeeze out under high axial loading (particularly if the wall is post-tensioned) and care should be taken to avoid these. Vertical d.p.c.s are, of course, required at openings as for traditional construction.

Temporary propping. Tall fin walls should have temporary props during the construction period before the roof is constructed and fixed.

Reinforced fin walls

Fin walls may be reinforced by building steel bars in the fins as the work proceeds or, when thermal insulation is not a consideration it can be achieved by alternative means, by grouting bars in the cavity. Reinforcement of specially designed hollow concrete blocks is another method which has proved to be highly successful with the blocks forming a permanent shutter for the grout or concrete.

Post-tensioned fin walls
Fin walls can readily be post-tensioned to avoid the development of high tensile bending stresses in the masonry. This technique involves bars being provided either in the cavity or in the fin wall during construction. The bars are anchored into a suitable foundation and tensioned via a capping plate at the top of the wall.

According to Curtin *et al.* (1984) it can be shown that the application of post-tensioning to plain masonry walls can usually increase the potential bending resistance by up to ten times. This also increases the section modulus over that of a comparable rectangular section by almost six times. Thus by using a post-tensioned fin wall, there is the possibility for an increase of some 60 times the moment of resistance of a comparable plain rectangular section. Post-tensioning is particularly suited to clay brickwork because of the high strength of many bricks together with their low creep characteristics.

Design of tall single-storey buildings

The roof

For the most economic design the roof must generally act as a plate or diaphragm to prop and tie the top of the walls and to transfer the resulting horizontal reactions to the end walls of the building (see Figure 12.6), assuming such transverse walls to be spaced within a reasonable span for the roof plate.

Alternatively, the roof decking may sometimes be used in conjunction with a concrete ring beam, or a wind girder can be used to provide horizontal roof bracing. The roof is usually supported on a concrete ring beam and this should be designed to

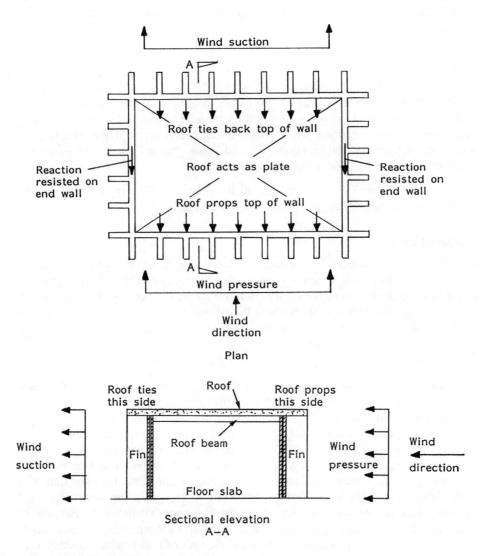

Figure 12.6 Wind loading and reactions

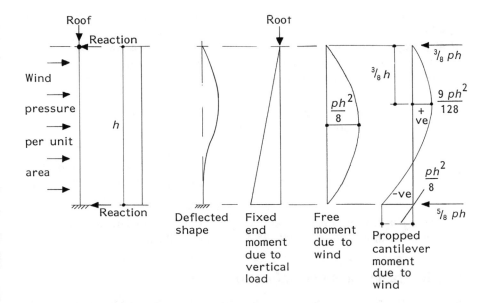

Figure 12.7   Structural behaviour of propped cantilevers

help transfer wind forces from the top of the walls on which the wind is acting to the end walls of the building. If capping beams are not used and the dead weight of the roof is small, the main roof beams often require to be tied down with anchors into the padstones and/or possibly into the fin wall below. It is important that capping beams do not bridge movement joints.

Whenever possible the support for the roof should coincide with the centroid of the fin walls to avoid unnecessary eccentricity of vertical loading on the walls.

Walls
The walls are normally designed as propped cantilevers. Figure 12.7 illustrates the propped cantilever condition for diaphragm walls which is equally relevant to fin walls. Alternatively, the fin walls can be designed as free cantilevers springing from foundation level.

Design procedure

(1) Determine the loading system for the building in the normal manner.
(2) Select a trial section.
(3) Calculate the size of the capping beam and/or alternatively determine the necessary anchorage for the roof.
(4) Check that the roof is capable of providing complete plate action or design a suitable bracing system.

(5) Calculate the bending moments and the stability moment at the base of the wall.
(6) Calculate the bending moment and moment of resistance at a position $\frac{3}{8}h$ from the top of the wall (assuming a propped cantilever design).
(7) Check the stresses for (5) and (6) above.
(8) Select a suitable unit and mortar strength.
(9) Calculate the shear stresses. If the fin is not fully bonded to the flange determine the type and spacing of the metal ties.
(10) Check the stability of the transverse walls for plate reactions (Figure 12.6).

For detailed examples of calculations and further design information readers are referred elsewhere (Curtin *et al.*, 1980, 1982, 1984; Curtin and Phipps, 1982; Phipps, 1987).

Materials components and workmanship. The materials components and workmanship for fin walls should comply with BS 5628: Part 3: 1985.

12.2 Diaphragm walls (unreinforced)

Traditionally, the roofs of most tall single-storey structures in sports and assembly halls, warehouses, theatres, gymnasiums, garages and churches are supported on steel or concrete columns.

Contractors and engineers are now using masonry diaphragm walls as an alternative form of construction. These generally produce a more elegant structural (and architectural) solution which is often more economic than the traditional solutions. Masonry diaphragm walls form the structure, cladding and lining in either brickwork or blockwork, use only one trade and can be insulated to any required level.

Diaphragm walls can also be free-standing boundary walls, low retaining walls and walls to carry heavy vertical loads. It is possible to increase the stability of boundary or retaining walls to resist lateral forces by filling the cavities with rubble or weak concrete. Foundations are also usually straightforward, only needing shallow strip footings. They avoid problems of differential movement.

A diaphragm wall has been defined (Phipps and Montague, 1976) as the wide cavity wall where the two leaves of the masonry are connected together not by cavity wall ties but by cross ribs of masonry. A diaphragm wall is stronger than a simple cavity wall because it no longer acts as two relatively flimsy leaves but as a series of connected box or T-sections (Figure 12.8), which have a high resistance to both vertical and horizontal loads.

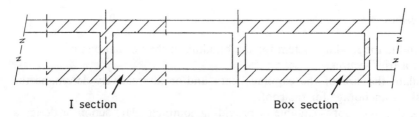

I section   Box section

Figure 12.8 Plan showing assumed I and box sections

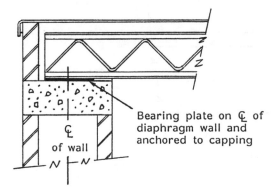

Figure 12.9   Roof bearing to ensure concentric loading and lateral prop to wall

If the roof is correctly seated on the top of the diaphragm wall (Figure 12.9), it acts as a lateral prop for the wall, thus reducing the wall thickness to a minimum.

Bonding of walls

The size of the units and joint thicknesses govern the plan dimensions of diaphragm walls. The designer is free to select the most appropriate shape to meet structural and architectural requirements. Units may be arranged in a number of different patterns (Figures 12.10a, b). Metal ties may be used instead of bricks or blocks to bond the cross ribs to the leaves (Figure 12.10c). Such ties are desirable where a d.p.c. is to be incorporated at the vertical joint between the ribs and the leaves. Plan profiles can be arranged so that some panels are recessed or protruding, or the ribs may project to provide an architectural feature.

With blockwork construction, special shaped units can produce alternative bonding arrangements such as quoin bonded walls. There must be provision for movement in long diaphragm walls as for traditional construction (see Chapter 4).

It may be necessary, for example, to provide vertical slip planes or movement joints at the corners of the building. In the length of the building, vertical movement joints can easily be accommodated at openings or by the provision of double ribs, one at each side of the joint (Figure 12.11).

To prevent problems of high local stresses due to horizontal wind loading and localized axial loads at beam bearings, extra ribs or thicker ribs each side of openings should be used to support the beams or lintels. As diaphragm walls have large voids it is possible to accommodate certain services within them. However, the designers and contractor must ensure that the location and size of any holes for services do not undermine the integrity of the walls. Under no circumstances chase walls for services.

The thermal transmission value (U-value) is similar to that of normal cavity walls but because air circulation can take place within the larger void and since the ribs create 'cold bridges' according to a BDA publication (Curtin and Shaw, 1977) the increase in U-value is estimated to be about 10 per cent. It is a simple matter to introduce insulating material into the voids for extra thermal insulation.

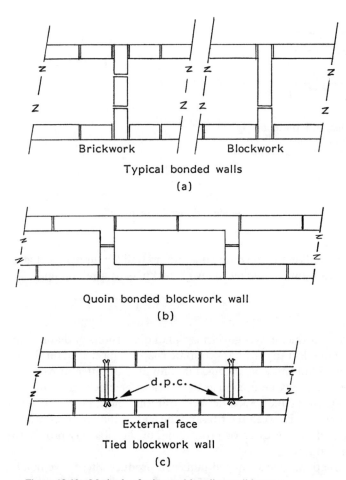

Brickwork

Blockwork

Typical bonded walls

(a)

Quoin bonded blockwork wall

(b)

d.p.c.

External face

Tied blockwork wall

(c)

Figure 12.10 Methods of tying and bonding wall leaves

Structural behaviour

The structural behaviour of diaphragm walls depends on the applied loading and support conditions. Free-standing walls will act as cantilevers which rely for their strength on the moment and shear capacity at the base or foundation. Diaphragm walls which support a roof capable of acting as a stiff horizontal plate may be assumed to act as propped cantilevers.

The loading on diaphragm walls is either vertical (self weight and roof loading) or horizontal due to wind forces or retained materials. Because of their geometrical properties, diaphragm walls have high axial load capacity and the greater the depth of the wall, the greater its resistance to lateral forces and the smaller its slenderness ratio.

To justify the propped cantilever, it is necessary to design the foundation to the wall to prevent failure due to overturning or sliding of the wall and the roof must be stiff enough to act as a prop capable of transferring all the horizontal forces. Figure 12.7 illustrates the propped cantilever condition and the resulting bending moments.

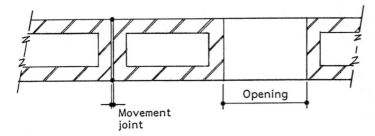

Figure 12.11   Provision for movement via movement joints or openings

According to Phipps and Montague (1976) there are four ways in which a diaphragm wall may fail in shear: (a) sliding at the base; (b) mortar joint cracking; (c) diagonal web cracking; and (d) for tied walls, tie failure (Figure 12.12).

Sliding at the base. This may be the critical feature for low retaining walls which are required to resist high lateral loads, particularly if they have a low-friction d.p.c. To avoid this problem, clay-brick diaphragm walls should have a brick d.p.c. and walls constructed of calcium silicate or concrete units should have a stepped d.p.c. (Figure 12.13).

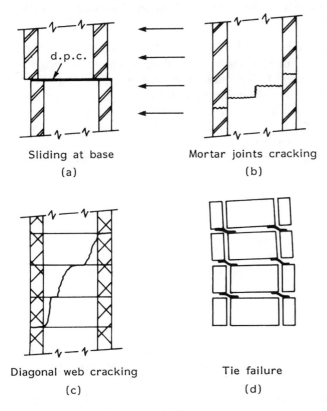

Figure 12.12   Possible modes of failure

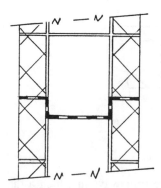

Figure 12.13   Stepped d.p.c

Mortar joint cracking. In the webs shear forces may be responsible for cracking both the vertical and horizontal mortar joints unless sufficient axial load is applied. If sufficient axial load does not come from the self weight of the construction, it may be necessary to impose higher axial forces by post-tensioning the walls.

Diagonal web cracking. When cracking of the mortar joints occurs in the webs this tends to start in the vertical joints and spread to the horizontal joints. It is therefore essential to ensure that all joints (particularly the perpends) are filled.

Tie failure. If there are wall ties, because the webs are not fully bonded to the leaves, the ties must be designed to transfer shear stresses. A damp proof membrane between the web and the external leaf should be satisfactory, providing that the wall ties are designed to resist the shear forces and the vertical joint containing the membrane is correctly filled with mortar so that the ties can develop their full strength. The ties should be of the corrosion-resistant metal strip type with a fish tail at each end. A single tie bridging the web from flange to flange is preferable to shorter ties at each end of the web. The ties should have a minimum embedment in either leaf of 50 mm.

Capping beams

Diaphragm walls should generally have a continuous reinforced capping beam, but the beam must not bridge any movement joints. A full width capping beam is preferable to one which sits only on the webs or on the web and one leaf. It is important that eccentric loading from roofs is not transmitted to capping beams.

Design procedure

(1)   Determine the loading system.
(2)   Select a trial section.
(3)   Calculate the size of the capping beam to resist wind uplift if it supports a roof.
(4)   Check that the roof is capable of providing complete plate action (if appropriate).
(5)   Calculate the bending moments and the stability moment at the base of the wall.

(6) Calculate bending moment and moment of resistance at $\frac{3}{8}h$ from top of wall (when it supports a roof).

(7) Check the stresses for (5) and (6) above.

(8) Select suitable unit and mortar.

(9) Calculate the shear stresses and the type and spacing of the metal ties.

(10) Check the stability of the transverse walls for plate reactions.

References

BS 5628: Part 3: 1985 *Use of masonry – materials and components, design and workmanship*. BSI, London.

Curtin, W. G. (1980) Design, theory and application of brick diaphragm walls. *The Structural Engineer*, Part A, February.

Curtin, W. G. and Phipps, M.E. (1982) *Post-tensioned brick diaphragm walls*. IBMAC.

Curtin, W. G. and Shaw, G. (1977) *Brick diaphragm walls in tall single-storey buildings*. BDA, November.

Curtin, W. G, Shaw, G, Beck, J. K. and Bray, W. A. (1982) *Design of brick diaphragm walls*. BDA, March.

Curtin, W. G. *et al.* (1980) *Design of brick fin walls in tall single-storey buildings*. BDA Publications, June.

Curtin, W. G. *et al.* (1982) *Structural masonry designers manual*. Granada Publishing Limited.

Curtin, W. G. *et al.* (1984) Masonry fin walls. *The Structural Engineer*, Vol. 62A, No. 7, July.

Korff, J. O. A. (1984) *Design of free-standing walls*. BDA Publication, February.

Morton, J. (1985) *Accidental damage robustness and stability*. BDA Publications, May.

Phipps, M. E. (1987) The design of slender masonry walls and columns of geometric cross-section to carry vertical load. *The Structural Engineer*, Vol. 65A, No. 12, December.

Phipps, M. E. and Montague, I. (1976) *The design of concrete blockwork diaphragm walls*. ACBA.

Roberts, J. J, Tovey, A. K, Cranston, W. B. and Beeby, A. W. (1982) *Concrete Masonry Designers Handbook*. A Viewpoint Publication.

13
Reinforced and post-tensioned walls

13.1 Reinforced masonry

Reinforced masonry has been used in this country for many years and records show that Mark Isambard Brunel used reinforced brickwork as long ago as 1825. Brunel not only reinforced vertical cylinders of brickwork as part of the Thames Tunnel Project, he also post-tensioned them.

The slow development of this form of construction may be attributed to lack of interest by architects, engineers and unit manufacturers and undoubtedly the availability of reinforced concrete with its plastic properties has not helped. Low permissible stresses in former British Codes to a large extent overrode such conditions; indeed they were perhaps the root cause.

The design of masonry generally is based on a gravity stable form of construction which allows little or no tension to develop. However, the introduction of reinforcement in the form of steel bars or mesh can produce an entirely new material and hence design concept. The use of reinforcement can be economically employed as follows:

(1) Vertical reinforcement to resist tensile forces due to wind loading.
(2) Light horizontal reinforcement in the bed joints, to increase the wall strength, especially under concentrated loads and where differential foundation settlement may occur.
(3) Light horizontal reinforcement in bed joints, to increase resistance to lateral loads.
(4) Light reinforcement horizontally and vertically in prefabricated panels, to resist handling stresses and to make walls span as deep beams or cantilevers.
(5) Horizontal or vertical steel reinforcement in special units to form beams or columns.
(6) Reinforcement in grouted cavity construction to provide additional vertical and lateral resistance to loading.

(7)   Pre-stressing wires vertically or horizontally, to provide pre-compression thereby reducing the effects of eccentricity of loading, etc. This latter form of construction will be discussed on p. 257.

Reinforced and post-tensioned masonry is used much more extensively in countries such as Japan, New Zealand and the USA where seismic forces need to be accommodated in designs.

Methods of reinforcing walls

In addition to placing light reinforcement in the joints the following methods are used:

(a)   Reinforcement incorporated vertically in pockets in one or both faces of a wall (Figure 13.1).
(b)   Grouted cavity construction using (i) mortared construction; (ii) short lift grouting or (iii) high lift grouting (Figure 13.2).
(c)   Quetta bond (Figure 13.3).
(d)   Reinforcing horizontally or vertically using special units (Figure 13.4).
(e)   Horizontally, to form beam sections (Figure 13.5).

Materials and components
Most materials used for the construction of unreinforced loadbearing masonry are suitable for reinforced masonry but BS 5628: Part 2: 1985 only gives design information for structural units having a compressive strength of 7 N/mm^2 or more. However, this does not imply that units having a lower compressive strength may not be used for certain applications. Indeed it is sometimes desirable to reinforce lightweight blockwork, for example, to control cracking (see Chapter 4).

Steel. Reinforcing steel is required to comply with the appropriate British Standards, i.e. BS 4449 (1978), 4461 (1978), 4482 (1969), 4483 (1969), and BS 970: Part 1 (1972). Special attention should be paid to durability when deciding upon the reinforcement to be specified.

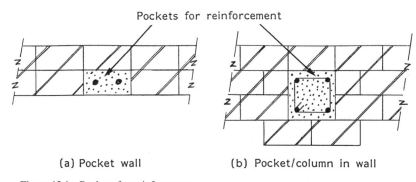

Pockets for reinforcement

(a) Pocket wall   (b) Pocket/column in wall

Figure 13.1   Pockets for reinforcement

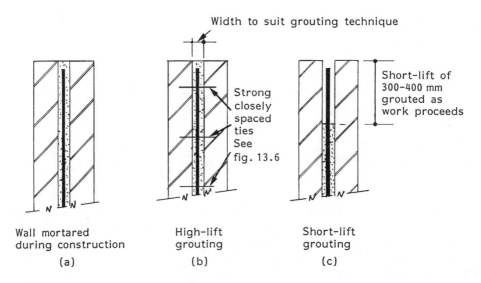

Figure 13.2   Grouted cavity walls

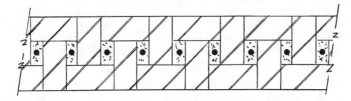

Quetta bond is similar to flemish bond but in plan the headers
and stretchers form a T rather than L shape. It facilitates the
introduction of vertical reinforcement in the voids, which are
then filled with mortar. The reinforcement is usually taken into
the foundations, floor and roof.

Figure 13.3   Quetta bond

Damp proof courses. Before selecting a d.p.c. care should be taken to ensure that it
will have the appropriate resistance to compression, tension, sliding and shear (see
Chapter 3).

Wall ties. The number and strength of wall ties for high-lift grouted construction
should be sufficient to resist the possible bursting forces. Figure 13.6 shows an
appropriate tying method. Vertical-twist ties to BS 1243 (1978) are recommended
for low-lift grouted cavity construction.

Mortar. Generally as for unreinforced masonry.

Concrete infill and grout. The Standard (BS 5628: Part 2: 1985) recommends a 1:0 to
$\frac{1}{4}$:3:2 Portland cement:lime:sand:10 mm maximum size aggregate or design mix of
grade 25 to BS 5328 (1981). The slump of such mixes should be appropriate to the
space to be filled, as well as the suction rate of the masonry units. The height to be

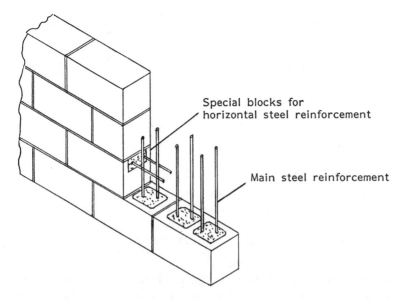

Figure 13.4   Reinforcing horizontally and vertically

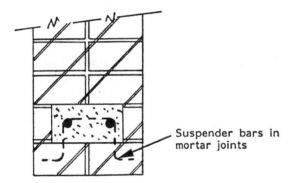

Figure 13.5   Wall beam section

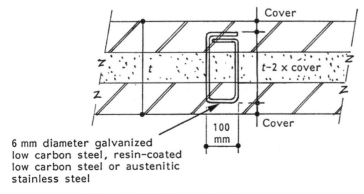

Figure 13.6   Wall ties for high-lift grouted cavity

filled is also important and in some areas plasticizers or super-plasticizers may be necessary.

Admixtures. Calcium chloride should never be used and other admixtures only with the permission of the designer.

Durability
The type of reinforcement and minimum level of protective coating is dependent upon the type of construction and the degree of exposure the masonry will be exposed to.

The Standard (BS 5628: Part 2: 1985) quotes four classifications of exposure and only allows the use of unprotected steel in grouted cavity or Quetta bond construction in Sheltered or Very Sheltered conditions. Elsewhere it requires the use of austenitic stainless steel or carbon steel galvanized (minimum zinc coating 940 g/m^2) or carbon steel coated with 1 mm minimum of stainless steel.

Effect of masonry units. According to the Standard (BS 5628: Part 2: 1985) 'The protection against corrosion provided by brickwork tends to be improved if high strength, low absorption bricks are used in strong mortar. Where bricks that have a greater water absorption than 10 per cent or concrete blocks having a net density less than 1500 kg/m^2, measured as described in BS 6073: Part 2 (1981), are used, the steel recommended for the next more severe exposure situation or, where appropriate, stainless steel should be used, unless protection to the reinforcement is to be provided by cover in accordance with the recommendations below'.

Cover to reinforcement. When austenitic stainless steel, or steel coated with at least 1 mm of austenitic stainless steel, is used there is no minimum cover required to ensure durability. However, some cover will be required if the full bond strength of the reinforcement is to be developed.

The minimum cover for reinforcement in bed joints is 15 mm and 20 mm for grouted-cavity construction or Quetta bond.

Wall ties. These should be specified so that their resistance to corrosion is at least equal to that of the reinforcement used in the same position, except that the minimum mass of zinc coating on galvanized steel ties should be 940 g/m^2. If the material used for the wall ties differs from that of the reinforcement, the two dissimilar metals should not be allowed to come into contact.

Structural design
BS 5628: Part 2: 1985 covers the design of reinforced masonry and is based on limit state principals in the same manner as plain masonry. The structural design of reinforced masonry is similar to the design of reinforced concrete and is beyond the scope of this book.

Readers are referred to the publications of the BDA and the CBA for detailed examples.

13.2 Post-tensioned masonry

It is not clear when post-tensioned masonry was first introduced but it was certainly used by Sir Mark Isambard Brunel (father of the great Isambard Kingdom Brunel) in connection with the Thames Tunnel in 1825. As a part of the construction of this tunnel, two brick shafts were built, each 762 mm thick, 15.240 m in diameter and 20.120 m deep. The shafts were reinforced (and post-tensioned) with wrought iron bolts 25.4 mm in diameter, built into the brickwork and attached to wooden curbs at the top and the bottom with nuts on the threaded wrought iron bars. Iron hoops 229 mm wide × 12.7 mm in thickness were laid in the mortar joints as the shafts were built. The first shaft was built to a height of 12.900 m and then sunk by excavating the earth from the interior, using what is now commonly known as the open method of cassion construction. The remaining 8.534 m of its height was added by underpinning. In spite of unequal settlement of the shaft as it was being sunk, no cracks developed in the masonry and as a result, the second shaft was built to its entire height of 20.120 m before being lowered.

Richard Beamis describes this construction and states: 'after an unequal settlement of 178 mm on one side and 89 mm on the other, the surge was alarming but so admirably was the structure bound (i.e. post-tensioned) together that no injury was sustained'.

The modern use of post-tensioned masonry has wide application. Typical examples are: reinforcement of traditional cavity walls; fin walls; diaphragm walls; special features such as clock towers; retaining walls and at least one example of a 'pre-stressed' brickwork water tank.

Principles of post-tensioned masonry

Plain masonry is relatively strong in compression (depending upon the type of unit and mortar used) but weak in tension. Therefore, when walls are subjected to bending either the thickness of the wall must be increased to produce zero or very little tension or the wall will require to be reinforced/post-tensioned. If reinforcement is introduced the tensile forces are taken by the steel reinforcement, whereas if an additional post-tensioning force is introduced this will have the effect of overcoming the tension introduced by the bending forces.

Figure 13.7 illustrates a wall concentrically loaded due to its self weight W_1 but eccentrically loaded by a force W_2 at a distance e from the centroid of the wall. In order to overcome the tensile bending forces the wall is post-tensioned with a further load W_3 (in this case positioned on the centre-line of the wall). The post-tensioning force is usually positioned on the centroid of the wall, which for unsymmetrical sections, will not coincide with the centre-line. The stress diagrams illustrate that as sufficient pre-compression (force W_3) has been applied, the summation of (a) + (b) + (c) gives compression at each side of the wall (diagram (d)). It is, of course, necessary to check that the wall is capable of sustaining the compressive forces applied to it.

The above is a simple example to illustrate the principles of design. However, this can be much more complex due to the properties of the materials involved which

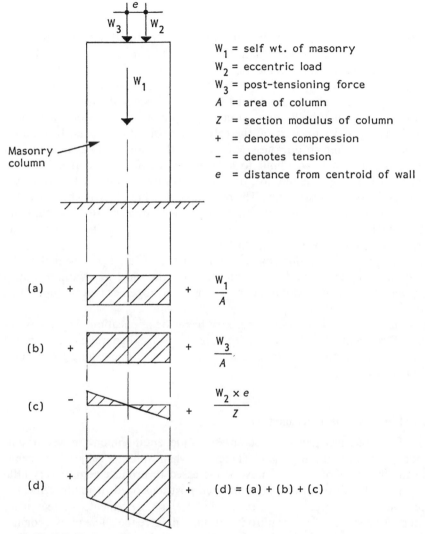

W_1 = self wt. of masonry

W_2 = eccentric load

W_3 = post-tensioning force

A = area of column

Z = section modulus of column

$+$ = denotes compression

$-$ = denotes tension

e = distance from centroid of wall

Figure 13.7 Pressure diagrams

may influence changes in the applied post-tensioning force after its application. The initial force may be reduced or increased due to some combination of the following:

(a) thermal movements
(b) drying shrinkage of the mortar and/or concrete units
(c) drying shrinkage of the mortar plus moisture expansion of clay units
(d) moisture movements in mortar and/or concrete units
(e) creep of the masonry
(f) elastic deformation of the masonry
(g) relaxation of the post-tensioning steel
(h) friction or slip.

According to Curtin *et al.* (1982) for most applications of post-tensioning to masonry it is considered that 20 per cent losses in the post-tensioning force due to all of the various factors may be assumed.

Losses of post-tensioning force

Creep and moisture movements. There is an instantaneous shortening of the masonry known as elastic deformation but a further time-dependent deformation also occurs and this is known as creep. The mechanism of creep differs dependent upon the type of units used. Clay masonry undergoes a negligible reversible movement due to wetting and drying. However, a much larger and irreversible expansion occurs due to moisture take up after removal of the units from the kiln. Most of this expansion occurs in the first few days but it can continue at a decreasing rate over a much longer period. Thus for clay brickwork the expansion of the bricks may almost equal the shrinkage of the mortar, resulting in a very small creep or shorting of the material. Conversely, concrete masonry does not grow with age but shrinks due to drying and carbonation shrinkage of the units as well as the mortar. Unlike clay masonry, concrete masonry is affected by wetting and drying. For more information on dimensional changes in the units readers are referred to Chapter 4.

Relaxation of post-tensioning steel. The amount of relaxation is dependent upon the type of steel, the degree to which the bars have been stressed and the period of time since stressing. It is therefore necessary to consider all these factors before making an allowance for relaxation of the bars.

Frictional losses. These occur mainly when stressing of the bars is carried out due to settlement of bearing plates, etc. on the masonry.

Method of construction

The foundations are cast in the normal way but special care is needed to ensure that the post-tensioning bars are positioned accurately in the foundation and the correct anchorage length provided. The wall is then built. If a solid wall, it will be necessary to leave holes around the bars for subsequent grouting; alternatively, when diaphragm or other wide cavity walls are post-tensioned the cavity is usually maintained. At the top of the wall, the bars are left with their threaded ends projecting (but protected) until the masonry has cured. A bearing plate is then slotted over the bars and secured with nuts. The nuts are tightened to the predetermined value using a calibrated torque spanner. An alternative is to use MacAlloy high tensile bars (with threaded coupler joints for long bars) and tensioning the bars is not by torquing but by using the standard jack for the purpose. If grouting of the pockets for the bars is carried out it is recommended that vent holes are included at stages throughout the height of the wall to prevent voids occurring in the slurry.

Where the post-tensioning bars are in unfilled cavities they should be suitably protected from corrosion. The current vogue is to wrap the bars with 'Denso' tape.

In high walls the post-tensioning bars need to be jointed. This is usually carried out using special couplers of adequate strength to resist the forces involved and these are also protected against corrosion. As for all high walls it is usually necessary to provide temporary props before tensioning the walls unless the wall is post-tensioned at various heights.

Provision for movement must be made in long post-tensioned walls as for traditional construction (see Chapter 4). However, as post-tensioning provides greater restraint (as for other types of highly stressed loadbearing masonry) the centres of movement joints can be further apart than for unrestrained walls.

Figure 13.8 illustrates a post-tensioned masonry retaining wall. In this example the bars are positioned eccentrically in the wall in order to counteract the overturning moment caused by the restrained earth. In multi-storey construction post-tensioning can be carried out in stages or the bars curtailed where smaller overturning forces occur (Figure 13.9).

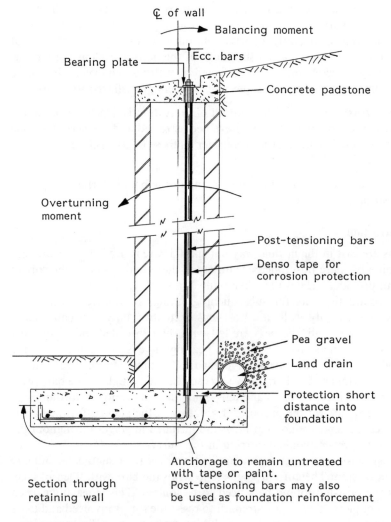

Figure 13.8 Post-tensioned retaining wall

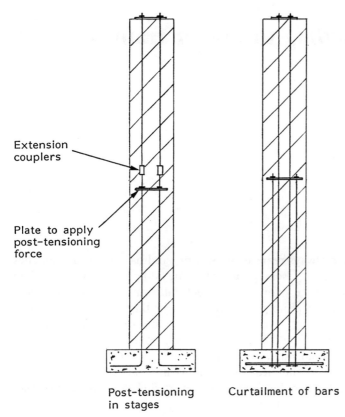

Extension
couplers

Plate to apply
post-tensioning
force

Post-tensioning   Curtailment of bars
in stages

Figure 13.9   Post-tensioning in stages and curtailment of bars

For examples of calculations readers are referred elsewhere (Curtin *et al.*, 1982; Allen, 1986; Shaw *et al.*, 1986).

References

Allen, L. N. (1986) *Post-tensioned brickwork at Rushden Fire Station*. BDA Engineers File Note, No. 1, March.

BS 5628: Part 2: 1985 *Use of masonry – structural use of reinforced and pre-stressed masonry*. BSI, London.

BS 4449: 1978 *Specification for hot rolled steel bars for the reinforcement of concrete*. BSI, London.

BS 4461: 1978 *Specification for cold worked steel bars for the reinforcement of concrete*. BSI, London.

BS 4482: 1969 *Hard drawn mild steel wire for the reinforcement of concrete*. BSI, London.

BS 4483: 1969 *Steel fabric for the reinforcement of concrete*. BSI, London.

BS 970: Part 1: 1972 *Wrought steels for mechanical and allied engineering purposes–general inspection and testing procedures and specific requirements for carbon, carbon manganese, alloy and stainless*. BSI, London.

BS 1243: 1978 *Specification for metal ties for cavity wall construction*. BSI, London.

BS 5328: 1981 *Methods for specifying concrete, including ready-mixed concrete*. BSI, London.

BS 6073: Part 2: 1981 *Method of specifying pre-cast concrete masonry units*. BSI, London.

Curtin, W. G. *et al.* (1982) *Structural masonry designers manual*. Granada Publishing Limited.

Shaw, G. *et al.* (1986) *The Osborn Memorial Halls at Boscombe*. BDA Engineers File Note, No. 6, November.

14
Masonry cladding to timber framed construction

Masonry and mortar have always represented security and have the characteristics most preferred by designers, the general public, planning authorities and the building societies for housing and indeed, many other structural forms. It is not surprising therefore, that the majority of timber framed houses in this country are clad in masonry, usually brick. Masonry provides the necessary non-combustible envelope which is aesthetically acceptable and combines this with excellent durability, good sound insulation, more than adequate structural strength, is virtually maintenance free and last but not least, economical. Masonry is not only a well-understood material by designers and builders but is also the preferred cladding for most structural frameworks, particularly housing.

14.1 The masonry

Masonry cladding or veneer (as it is sometimes referred to by the timber frame industry) has to fulfil numerous functions and should be constructed in accordance with BS 5628: Part 3: 1985. Until recently it has been assumed that the timber framework carried all the vertical and horizontal loads and the masonry merely provided the preferred envelope to satisfy UK tastes. However, recent research has shown that the masonry can, if connected to the timber frame with suitable ties, provide considerable stiffness to the construction. This is particularly so if the masonry is full height and has returns. BS 5628: Part 6: 1988 provides design guidance on the contribution of masonry 'veneer' to racking resistance.

There are potential dangers if racking resistance is assumed with insufficiently stiff connections to the timber frame. Conversely, if rigid or non-flexible ties are used cracking of the masonry may result. Current thinking on the maximum contribution to the permissible racking resistance provided by the masonry limits it to 20 per cent of the permissible racking resistance provided by the timber framed wall when considering wind forces in any one direction.

As for traditional construction all units should be laid on a full bed of mortar and if frogged bricks are used it is recommended that for optimum performance the units should be laid frog-up and all the frogs filled with mortar. It is also important to ensure that all perpend joints are solidly filled with mortar other than those specifically left open as weep holes.

The faces of joints can be worked to a profile or flush-finished. Recessed joints are not generally recommended by the writer for single-leaf masonry when used as cladding, but if the designer wishes to use this type of joint it is strongly recommended that the work is continuously supervised during construction to ensure a high standard of workmanship.

If coloured mortar is specified it is suggested that the whole bed should be coloured mortar and not just the pointing. However, if pointing with a coloured mortar is carried out it is important that the strength of the pointing matches that of the bedding mortar, bearing in mind that additions of pigment (particularly carbon black) often reduce the strength and consequently the bond between the unit and mortar.

Some lightweight units may not be satisfactory for use in single-leaf external work; if in doubt always consult the manufacturer. For suitable mortar designations specifiers should consult Chapter 3 or BS 5628: Part 3: 1985.

Movement

When designing timber framed construction it is currently assumed that the framework takes all the load (both wind loading and dead/superimposed vertical loading) and that the masonry cladding merely transmits the wind loading via the flexible wall ties to the frame. The actual amount of lateral load carried by the masonry depends on several factors, particularly the returns at corners.

Regardless of whether any account is taken of the contribution to lateral strength, as discussed earlier, accidental loading of the masonry due to differential movement of the component materials should be avoided to prevent random concentrations of stress on the masonry and subsequent problems of cracking.

If clay masonry is used this tends to expand due to thermal and moisture movement, whereas concrete and calcium silicate units tend to contract (see Chapter 4). However, in timber frame construction the largest movements take place due to the drying of the timber, the effect of compressive loading on the frame joints and long-term deformation of the timber (creep).

Shrinkage occurs almost entirely in the horizontal components of the frame and the total movement could be as much as 10 mm at the eaves level of a two-storey house and 15 mm in a three-storey house. Clay masonry tends to expand slightly due to moisture/thermal effects resulting in a total movement of approximately 5 mm at the eaves of a two-storey house and 7.5 mm in a three-storey house. Conversely, calcium silicate and concrete masonry shrinks slightly to approximately the same order of magnitude.

It is therefore necessary to make provision for movement at each location where an element supported by the timber framework bridges the masonry, e.g. at eaves, verges and cills. The recommended allowance for movement gaps to accommodate relative masonry and timber moisture and thermal movements is:

	Clay	Concrete or calcium silicate
Up to first floor	10 mm	5 mm
Up to second floor	15 mm	5 mm
Up to third floor	20 mm	10 mm

In traditional masonry construction lintels frequently bridge the cavity wall without movement problems, but in timber frame construction due to the differing movement characteristics it is necessary to separate lintels for the masonry from the timber frame. Similarly, door and window frames should not be fixed to the masonry although this is unlikely to be a critical factor at ground floor level.

Suitable fillers for movement gaps at eaves, verges, door and window frames are hand-applied bedding compounds or butyl/rubber and polyisobutylene strip. Cellular polyethylene or cellular polyurethane may be suitable and are also appropriate for movement joints in the masonry.

Materials recommended for sealing joints around doors and window frames are gun-applied, non-curing oil-based or acrylic sealants, or gun-applied one- or two-part chemical curing polysulphide or silicone sealants.

Movement gaps between the top of the masonry and the underside of the roof structure need not be filled provided the joint is adequately protected by the eaves overhang (Figure 14.1). If it is considered desirable to fill the gap and the roof space is correctly ventilated, then a compressible jointing material such as bitumen-impregnated polyurethane may be used. Manufacturers should always be consulted on the suitability of their materials for the proposed use and the joints made in accordance with their recommendations. It is also recommended that vertical movement joints be provided in long lengths of masonry to accommodate horizontal movement of the cladding. For clay brickwork 10 mm wide joints at up to 12 m centres are generally recommended and short returns (less than 686 mm) should be avoided in long lengths unless a vertical slip plane is provided to allow for movement, see (Chapter 4). Similar provision for horizontal movement should be made in calcium silicate and concrete units at maximum centres of 7.5 – 9 m and 6 m respectively. BS 5628: Part 3: 1985 gives more information on movement joints and the reader is also referred to BRE Digest No. 137 (1972) for joint design.

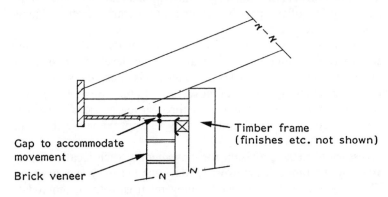

Gap to accommodate movement

Brick veneer

Timber frame
(finishes etc. not shown)

Figure 14.1   Eaves overhang

Wall ties

There is currently no British Standard for the type of flexible wall ties (and fixings) recommended for use in conjunction with masonry clad timber frame construction but NHBC Practice Note 5 (1982) 'Timber framed dwellings' specifies that they should be:

(a)   flexible to allow differential movement;
(b)   manufactured from acceptable materials such as austenitic stainless steel, phosphor bronze or silicon bronze; and
(c)   made from materials which are compatible with each other in order to avoid the risk of electrolytic corrosion. Any combination of stainless steel, phosphor bronze and silicon bronze being acceptable to the NHBC.

Less durable wall ties may be acceptable by some authorities but the author would strongly recommend the NHBC specification.

The flexible wall ties are normally located on the centre of the studs 400–600 mm horizontally and every sixth course vertically (450 mm) and staggered. It is absolutely essential that the nails securing the wall ties are in accordance with the tie manufacturer's specification and that they are located accurately through the sheathing into the studs. Nails should never be omitted and it is unacceptable to rely on fixing only to the structural sheathing. Generally, two nails are used to fix each tie and bricklayers should not omit one of the two nails unless specifically instructed to do so by the designer. Some ties are designed to be used with only one (ring shank/improved) nail which may be satisfactory if used at the centres specified for the product. Indeed, the Building Research Establishment took the initiative in proposing a performance standard for wall ties and in their publication (de Vekey, 1984) advocated the use of a single-hole fixing.

Additional ties should be placed every 300 mm vertically at openings and at both sides of corners. The wall ties should project at least 50 mm into and be properly bedded in, the horizontal mortar joints. They should be installed with a marked slope down to the exterior and when Chevron-type ties are used the arrows should point upwards to conduct any moisture to the outside. It is important that some indication is provided on the breather paper as to the location of the centre of the studs in order that the bricklayer/mason can nail the ties in a satisfactory manner. For guidance on testing and design of wall ties readers are referred to DD 140: Part 1 (1986) and Part 2 (1987).

Cavity barriers

Because the cavities in timber frame construction contain combustible materials, the Building Regulations require cavity barriers to be incorporated vertically at intervals not exceeding 8 m horizontally, at separating wall/external wall junctions, around all openings and at floor/wall and roof/wall junctions. Cavity barriers may be formed of:

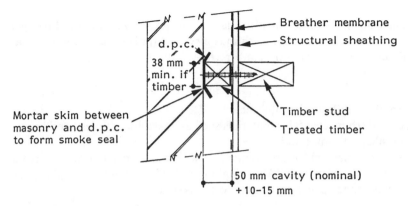

Figure 14.2   Cavity barriers

(a)   Preservatively treated timber at least 38 mm thick secured to the timber frame. This construction requires considerable care in order to ensure the required sealing of the cavity between the frame and the inner face of the masonry. It is also essential that a d.p.c. is provided between the cavity barrier and the masonry (Figure 14.2). It has the advantage that it also provides a positive support for the masonry laterally against wind pressure.

(b)   Mineral fibre enclosed in polythene may be used which, due to its resilience, can provide a more positive seal between the masonry and the timber frame with less difficulty than the method described above.

Breather papers

The function of the breather paper is to provide temporary protection against rain ingress to the timber frame walling during construction before the masonry is complete. It also provides permanent resistance to any small amount of water which might accidentally penetrate the cladding and cross the cavity during freak conditions. While doing this, it must not materially impede the passage of water vapour from the interior of the dwelling to the outside. It is important that breather papers have adequate wet strength to resist mechanical and wind damage while exposed to the elements.

NHBC require breather membranes to have a vapour resistance less than 0.6 MNs/g when calculated from the results of tests carried out in accordance with BS 3177 (1959) at 25°C and relative humidity of 75 per cent. It is important to ensure that the breather paper is adequately lapped 100 mm on horizontal laps and 150 mm on vertical laps, fixing the upper layers over the lower layers to ensure that the rainwater runs away from the sheathing.

Damp proof course (d.p.c.)

The d.p.c. in the outer leaf should comply with the appropriate British Standard or Agrément Certificate and be of the flexible type. It should be bedded above and below in mortar and properly lapped at least 100 mm at joints. The d.p.c. should be located at least 150 mm above finished ground floor level.

Lintels

If proprietary lintels supporting the outer leaf of masonry bridge the cavity, they must not be fixed rigidly to the timber frame. When clips are provided they should be secured at 600 mm intervals with provision for movement.

Proprietary lintels should always be installed in accordance with the manufacturer's recommendations. When angle lintels are used they should be bedded level. It may be preferable to use angle lintels over the top floor openings to enable the top courses of masonry to be tied back to the structure. This may not be possible with proprietary lintels.

Cavity trays

Cavity trays are normally provided over openings and horizontal cavity fire barriers unless the specific area is protected by a roof projection. They should be carried up across the cavity, lapped under the breather paper weathering and secured to the timber frame and extend through to the face of the masonry cladding. Cavity trays should be one piece material and carefully installed to prevent puncturing or tearing.

Weep holes

Weep holes should be provided via the perpend joints in the masonry cladding below the d.p.c. (Figure 14.3) and immediately above horizontal cavity barriers and cavity trays over lintels. They should be located at 900 mm intervals thus providing draining and ventilation of the cavity.

Cavities and tolerances

In the early days of timber frame construction in the UK small cavities were permitted between the inner face of the external masonry and the structural sheathing of the timber frame. NHBC now require a 50 mm nominal ventilated cavity with a tolerance of + 10 mm – 15 mm. However, if timber cavity barriers are used requiring

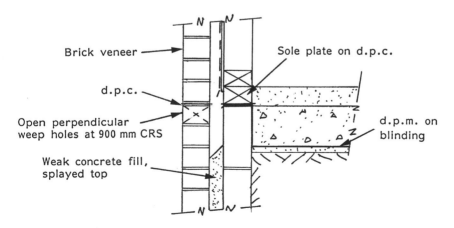

Figure 14.3   Section through wall at ground floor level

a tight fit this will call for a disciplined approach to the setting out and bonding of the masonry, which should be given due consideration by all concerned and may call for closer tolerances than specified above.

Wall insulation

Insulation of the external wall should be installed between the structural sheathing and the dry lining of the timber frame. Insulation should not, under any circumstances, be placed in the cavity between the masonry cladding and the structural sheathing either during construction or on any subsequent occasion.

Further information

For further information on brick cladding to timber frame construction readers are referred to BDA Design Note 6 (1982).

References

BDA Design Note 6 (1982) *Brick cladding to timber frame construction.* March.
BS 5628: Part 3: 1985 *Use of masonry – materials and components, design and workmanship.* BSI, London.
BS 5628: Part 6: 1988 *Code of Practice for timber frame walls Section 6.1 Dwellings not exceeding three storeys.* BSI, London.
BS 3177: 1959 *Method for determining the permeability to water vapour of flexible sheet materials used for packaging.* BSI, London.
Buildings Research Establishment Digest 137 (1972) *Principles of joint design.* January.
DD 140: Part 1: 1986 *Wall ties – methods of test for mortar joint and timber frame connections.* BSI, London
DD 140: Part 2: 1987 *Wall ties – recommendations for design of wall ties.* BSI, London
de Vekey, R. C. (1984) *Performance specifications for wall ties.* BRE
NHBC Practice Note 5 (1982) *Timber framed dwellings and external timber framed wall panels in masonry cross wall dwellings.* NHBC.

According to BRE Digest 273: *Perforated bricks* (1983) there is not much difference in the performance of solid bricks and those having up to 25 per cent perforations, although one slotted type appears to be more vulnerable than the others. There is definite evidence that bricks with higher levels of perforation (above 25 per cent) fail by a different mechanism from solid bricks and at markedly shorter times. However, few if any bricks with greater than 25 per cent perforations are produced currently. If designers are in doubt they should request test data from the brick manufacturers.

Concrete
Concrete blockwork produced using limestone, air-cooled blast-furnace slag, foamed or expanded slag, crushed brick, well-burnt clinker, expanded clay or shale, sintered pulverized-fuel ash and pumice are given much higher notional fire resistance ratings than those produced using aggregates based on gravels and crushed natural stone, with the exception of limestone. Concrete blockwork made from lightweight aggregates and calcareous aggregates are much less likely to spall in fires than the siliceous-aggregate concrete units.

Units made from autoclaved aerated concrete behave in a different manner to some of the other lightweight concrete units when loadbearing and under fire test and consequently have a different notional fire resistance for the longer tests. Nevertheless, autoclaved aerated concrete walls still achieve satisfactory fire resistance.

Calcium silicate bricks
These behave in a similar manner to concrete bricks and BS 5628: Part 3: 1985 gives both units the same notional fire rating.

Plaster
BS 5628: Part 3: 1985 gives fire ratings for walls plastered or rendered with not less than 13 mm of sand:cement or sand:gypsum (with or without lime) also vermiculite:gypsum plaster ($1\frac{1}{2}$:1 to 2:1 by volume) and perlite may be substituted for vermiculite for fired-clay bricks. Plasterboard is also permitted as an alternative to sand:cement or sand:gypsum for fire resistance periods up to 2 hours. Plasters are non-combustible when tested in accordance with BS 476: Part 4: 1970 and do not contribute to the spread of flame over the surface of a wall unless covered with a combustible finish (such as wallpaper). Plasterboard on the other hand is combustible (due to the paper liners). However, many authorities including the Fire Protection Association regard plasterboard as being equivalent to a non-combustible material.

Movement joints
Where movement joints or edge clearances are required for walls designed to resist fire, the filler material should be non-combustible, e.g. mineral fibre, and be capable of allowing the movement joint to function efficiently. Consideration should also be given to non-combustible cover strips fixed to both faces of the wall on one side of the joint.

15.3   Effects of fire on mortar and concrete

The BRE (HMSO National Building Studies, 1949) carried out a survey of fire-damaged buildings during the winter of 1941 at a time when ample data were available as a result of enemy attack. The survey provided much valuable information and it is now possible to estimate the temperatures of certain fires and to note if permanent damage has occurred to mortars and concretes.

Experiments carried out after the survey provided the following information:

(a)   The development of a red or pink colouration in concrete or mortar containing natural sands or aggregates of appreciable iron oxide content occurs at 250–300°C. The demarcation between the changed and unchanged material is usually sharp and the depth which has been heated above 300°C can generally be judged to ± 3 mm if a good section is available.

(b)   The second definite change or series of changes occur around 600°C with siliceous aggregates but is less sharp than at the 300°C change. The disappearance of the red or pink colour with a return to a grey occurs generally between 600°C and 700°C. Expansion effects such as cracking of flint gravels and general weakening or friability of the concrete or mortar are evident at 500–600°C.

(c)   A change from the second grey colour to a buff is often rather ill-defined and the temperature varies with the rate of heating in the range 900–1500°C.

(d)   Sintering may occur in prolonged fires when the temperature approaches 1200°C.

15.4   Effects of fire on wall ties

Little information is available on the fire resistance of wall ties or indeed on the effect when ties may be damaged after major fires. It is suggested that when cavity walls are subjected to major fires some check on the ties is necessary before passing a wall as sound, based merely on external visual inspection. The reason for this recommendation is that in at least one instance following fire damage, corroded vertical twist ties have been responsible for structural distress (zinc vaporizes at approximately 400°C).

Similarly, if polypropylene wall ties have been used these should also be checked to ensure that they have not deteriorated.

15.5   Effect of fire on damp proof courses

Damp proof courses and other built-in components should be checked after fires as these may be damaged and/or rendered ineffective.

References

BS 476: Part 8: 1972 *Test methods and criteria for the fire resistance of elements of building construction.* BSI, London.
BS 5628: Part 3: 1985 *Use of masonry – materials and components, design and workmanship.* BSI, London.
BS 476: Part 4: 1970 *Non-combustibility test for materials.* BSI, London.

Fire resi.

Buildings Research Establishment Digest 273 *Perforated bricks.*
CIRIA Technical Note 118. *Spalling of Concrete in Fires.* 1984
HMSO Notional Building Studies Technical Paper No. 4. *Investigations on Building Fires Part 1 Estimation of Maximum Temperatures Part II Colour Changes in Concrete or Mortar.*
The Building Regulations – Approved Document B. *Fire Spread.* HMSO, 1985

ınship, quality control, bonds and

16.1 Workmanship and quality control

Perhaps the one most critical factor affecting the strength, resistance to rain penetration, sound insulation and aesthetics of masonry is the quality of workmanship. There can be little doubt that the quality of work now accepted as normal falls far below the recommendations of the relevant British Standards, which can readily be achieved by competent tradesmen.

The term 'workmanship' not only applies to the care and technique with which the units are laid but also to certain factors that are in effect part of the wall design and specifications but which depend on the mason or bricklayer for proper execution.

Workmanship as a function of materials

Workmanship is not only the technique of laying the units but also that of selecting the correct units and ensuring that the appropriate mortar is used. Some authorities claim that the most important function in the bricklaying or blocklaying process (after ensuring that the correct mortar mix is used) is to ensure that the correct workability of the mortar is achieved as an aid in filling the joints. Experienced tradesmen will ensure that the mortar has a high water retentivity for units of medium or high absorption and a drier mix for low-suction units.

Generally, mortars of low water retentivity have harsh working properties whereas mortars high in water retentivity are plastic and easily worked. Some bricks (but not concrete/calcium silicate bricks or blocks) may need to have their suction rate reduced by immersion in water but this needs to be controlled and hosing down in hot weather is not recommended as this tends to saturate the external units in the pack or stack and leave those in the centre bone dry.

Filling of joints. Masonry units are frequently laid with only the outer edge of the vertical (perpend) joints 'buttered' with a small quantity of mortar so that within the wall they remain unfilled. The specification should require all perpend joints to be filled. However, the degree of filling of vertical joints depends largely on the care and skill of the bricklayer or blocklayer and on small projects this is normally left entirely to the craftsman.

Incomplete filling of perpend joints may not affect the vertical loadbearing capacity of masonry but it certainly reduces the lateral strength and resistance to rain penetration, etc.

Furrowing of mortar beds. One technique used by bricklayers and blocklayers is to lay a ribbon of mortar and then to furrow it with the point of the trowel. This is carried out to speed the work and also in some instances to make it easier to settle the units to line and level. This technique is not recommended as it: (a) makes the wall less resistant to compression; (b) produces reduced tensile bond, i.e. adhesion between the mortar and the units; and (c) makes the wall less resistant to rain penetration.

Many authorities agree that the joints in a wall should be solidly filled to achieve optimum performance. However, some block manufacturer's recommend that their products be laid on strips of mortar to improve thermal insulation. This technique can be successfully used but does include the penalties mentioned above. Designers should decide on their priorities and specify accordingly.

Joint thickness. It is generally assumed that a 10 mm joint will be used with standard bricks and blocks. Excessively thick joints tend to require a stiff, harsh-working mortar and very little pressure may be applied to the mortar. This results in poor contact and a less watertight joint; thick joints also tend to result in walls of lower compressive and tensile strength. Conversely, excessively thin joints are equally unsatisfactory as they require considerable force to place the unit in the intended position and unless the units have an extremely even bed face the mortar is incapable of tolerating oversize aggregate in the mortar or high points on the units without the danger of high local stresses occurring.

Consequently the bricklayer or blocklayer must use the joint thickness specified (usually 10 mm) to ensure optimum performance. However, when building masonry into framed construction the joint thickness may need to be modified slightly to accommodate building tolerances.

Joint profiles. Well-formed joint profiles enhance the appearance of masonry and have a considerable effect upon the resistance to rain penetration and the structural properties of walls. Concave tooled joints give the greatest resistance to rain penetration but the joint tooling is not as important as the workmanship inside the wall. Flat-tooled, rough-cut or raked mortar joints tend to produce walls with less resistance to rain penetration.

Tooling of the joints tends to assist in bringing the mortar and units into intimate contact but should not be carried out before the mortar has stiffened enough to hold its shape. Premature tooling of the joints can be detrimental to bond strength.

Recessed joints should only be used with bricks of proven frost resistance because of the danger of spalling of the edges and never when bricks have a row of vertical perforations near to the outer edge, otherwise there is a danger of rain collecting there. Slow seepage from such reservoirs so formed can disfigure the brickwork with efflorescence or free lime staining from the mortar.

Specifiers should also bear in mind that deeply raked joints markedly reduce the compressive strength of masonry.

Frogged bricks. In loadbearing masonry deeply frogged bricks should be laid frog-up and the frogs should be filled with mortar to obtain optimum loadbearing capacity. However, the BRE (1956) have shown that for normal two-storey domestic construction deep-frogged bricks can be laid frog-down and achieve a satisfactory performance with good quality workmanship.

Surface texture. Research has shown that for units having a rough texture not all the small surface indentations are likely to be filled unless the mortar is very workable. Even with the most workable mortars if the unit has a high suction rate the mortar is likely to stiffen too rapidly to fill the voids. Dusty bricks and bricks with sand from the moulds adhering to them may prevent proper contact between mortar and bricks. Such bricks should have all dust and sand on the bed faces removed from them before bricklaying begins.

Colour variation. To avoid a patchy appearance care should be taken to mix facing masonry units from different consignments. Colour variation in different batches of mortar, which will also lead to uneven appearance, may be reduced by consistent mixing and preparation.

When laying masonry units the mortar should not be allowed to come into contact with their faces since removal of mortar stains is not a simple matter. This applies particularly to open-textured units. Whenever practicable, facing work should not be racked back and left overnight before being brought level. The appearance of finished masonry will undoubtedly be affected by failure to protect the work during construction. In multi-storey construction the scaffolding boards adjacent to the work should be turned back at night to avoid splashing during wet weather.

Setting out

When setting out masonry care should be taken to reduce the cutting of units to a minimum and to avoid irregular or broken bond, especially at openings or in piers. Particular care should be exercised in the accuracy of setting out of the first course of masonry units in order to avoid subsequent inaccuracies in the finished work.

The horizontal distance between cross joints in successive courses should normally be not less than one-quarter of the length of the units. Where corners and other advanced work are raised above the general level they should be racked back not higher than 1.2 m at one lift and for facing work the whole lift completed in one operation.

Mixing and use of mortars

Mortar made on site should preferably be mixed by machine. The mixer and platform should be cleaned before use, when changing mixes, particularly coloured mortars and immediately after mixing is completed. When mixing, care should be taken to ensure that the correct quantity of water is used, as too much will produce light shades of mortar, reduced strength and durability.

Consistency, both as regards proportioning of materials and mixing time, is essential to produce good mortar and consistent masonry. Wide variation in mixing time should be avoided, particularly when plasticizers are added to the mix. In general a mixing time of between 3 and 5 minutes after all the constituents have been added to the mixer should be satisfactory. When the mortar is being made from coarse stuff (lime:sand), about three-quarters of the required mixing water should be added to the mixer followed by the required quantity of cement, which should be added slowly to ensure a thin paste free from lumps. The required quantity of coarse stuff should then be added and allowed to mix in together with any additional water to achieve workability.

Special cement, retarded ready-to-use cement:lime:sand and cement:sand mortars and dry-packaged cementitious mixes should only be used when specified and then strictly in accordance with the manufacturer's instructions.

Admixtures added on site should be used in accordance with the manufacturer's instructions and only with the permission of the specifier.

The materials for mortar should be measured accurately using a gauge box, bucket or similar container of known volume unless weight batching is specified. Gauging by shovels can give inconsistent results.

Wall ties

In cavity walls the leaves should be united by wall ties embedded in the mortar with a slight slope downwards towards the outer leaf. The ties should be positioned at the time the appropriate masonry courses are laid to a minimum depth of 50 mm. The ties should be placed at a rate of not less than 2.5 per square metre and should be staggered and evenly distributed. For example, in non-loadbearing masonry they should normally be at intervals of not more than 900 mm horizontally and not more than 450 mm vertically; in loadbearing masonry the spacing may need to be at closer centres (see BS 5628: Part 1: 1978). Additional ties should also be provided near the sides of all openings so that there is one for each 300 mm of height of the opening.

Wall ties should never be pushed into mortar joints after the masonry has been built, nor should they be placed in perpend joints. Rain penetration of masonry frequently occurs when wall ties are bridged with mortar and to prevent this happening laths should be placed in the cavity and subsequently drawn up as the work proceeds.

Cavity trays and d.p.c.s

All cavity trays and horizontal d.p.c.s should be laid within a bed of mortar across the full width of the external leaf of masonry and preferably projecting beyond the external face to form a drip. The material forming the damp proof membrane should never be inset from the face of the masonry and pointed over as this allows rising damp to bypass the d.p.c. and may cause cracking of the masonry if differential movement takes place between the pointing and a less rigid membrane.

To ensure an effective d.p.c. the recommendations on p.118 should be followed and special care taken with flexible d.p.c.s during frosty weather. Cracking of the membrane may be caused by bending over sharp edges and careless bedding on oversized aggregates or penetration by the point of a trowel can render the d.p.c. totally ineffective.

Structural masonry

Structural masonry needs a higher level of quality control than non-loadbearing domestic construction and normal practices of forming chases and holes, etc. may not be acceptable. Specifiers are referred to the Model Specification for Clay and Calcium Silicate Structural Brickwork (British Ceramic Research Ltd., 1988) and BS 5628: Parts 1 (1978), 2 (1985) amd 3 (1985) for further detailed information. Control testing on site and permanent site supervision are frequently necessary for this specialized type of work.

16.2   Blockwork bond patterns

General considerations

The choice of bond may be dependent upon strength requirements and/or aesthetics and the bond patterns illustrated are only a few of the most commonly used. Interesting effects can be achieved by projecting some of the units in a wall as it is constructed so that they project on one side and form a recess on the other face, or by the use of deeper blocks which can be laid to project on one or both faces of the wall.

When selecting a bond the stability of the wall must be considered and normally the bed joints should be arranged so that at least a quarter of the length of the units overlap. For maximum bond strength the overlap should be half the length of the unit.

It is important when using patterns which do not meet the above criterion to ensure that stability is achieved by alternative methods of support and/or by the use of suitable reinforcement.

In addition to aesthetic considerations bond patterns should facilitate the distribution of loads uniformly throughout the wall and accommodate, to some extent, building movement.

Bond types

Running Bond. This is the most commonly used bond for concrete masonry as it provides loadbearing walls of optimum strength and all round stability (Figure 16.1). It also provides good flexural strength horizontally.

Stack Bond. Stack bond (Figures 16.2, 16.3 and 16.8) is the quickest and most economical bond to lay as it eliminates the need for half-blocks (unless half-length stack bond is used) and reduces the need for cutting or the provision of special units. Its distinctive uniform pattern is particularly suitable for infill panels in framed structures as well as for non-loadbearing partitions. In most panels it will be necessary to reinforce alternate bed joints (or vertical joints if more appropriate) to achieve stability. Interesting patterns can also be achieved by varying course heights or by using blocks of different dimensions.

Figure 16.1   Blockwork – running bond

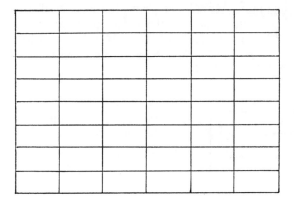

Figure 16.2   Blockwork – horizontal stack bond

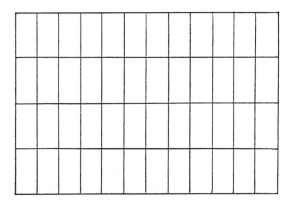

Figure 16.3   Blockwork – vertical stack bond

Basket Weave. Various intricate bonds can be achieved using a mixture of horizontal and vertical blocks of the same size or incorporating a range of sizes (Figures 16.4, 16.5 and 16.9). In order to avoid problems with bonding and cutting it is advisable to use only solid blocks for this type of bond.

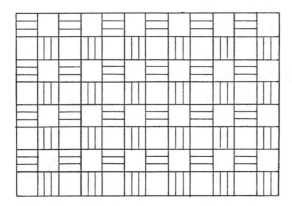

Figure 16.4   Block/brick – basket weave bond

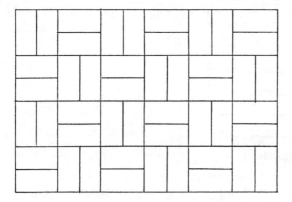

Figure 16.5   Blockwork – basket weave bond

Ashlar bond. Ashlar bond (Figure 16.7) can be achieved by the use of alternative courses of concrete blocks and concrete bricks. Coursed Ashlar alternate courses (Figure 16.10) is achieved in a similar manner by the use of two courses of blockwork to one course of concrete bricks. Interesting effects can be achieved by the use of bricks having contrasting colours to the concrete blocks.

Special bonds. Special bonds (Figures 16.6 and 16.11) can be achieved using concrete bricks and blocks of varying dimensions or by laying the blocks to an incline. However, such practices, while producing interesting results, may prove to be expensive as the walls often need to be contained or reinforced for stability, require several types of unit and more skilled workmanship than for simple bonds.

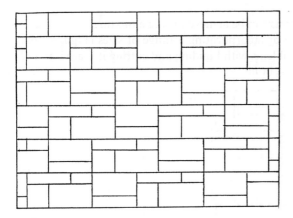

Figure 16.6   Block/brick – a special bond

Figure 16.7   Block/brick – Ashlar bond

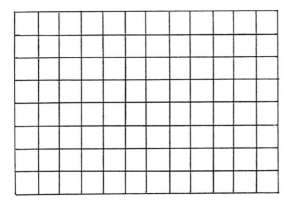

Figure 16.8   Blockwork – half-length stack bond

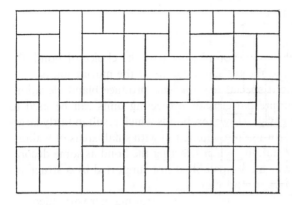

Figure 16.9   Blockwork – basket weave bond

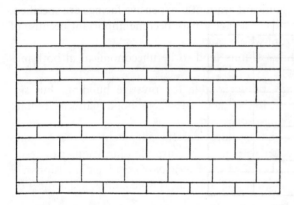

Figure 16.10   Block/brick – coursed Ashlar (alternate courses)

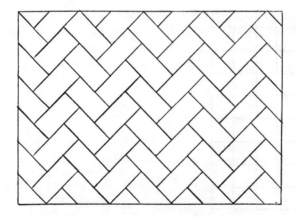

Figure 16.11   Blockwork – herringbone pattern

16.3   Brickwork bond patterns

General considerations

The appearance of facing brickwork can be influenced by the choice of bond, the colour and texture of the bricks and the type and colour of the mortar joints.

Large areas of brickwork without special features can produce bland elevations and to avoid such monotony definite patterns or projecting units can be used to enhance an otherwise uninteresting wall. However, bonds which result in fairly large-scale patterns may not be suitable for use in conjunction with small areas of walling. It is always important to consider stability when selecting the bond as some decorative patterns may result in walls which are unstable if not supported by a structural framework or tied to a suitable background material.

To ensure good bond the bricks must be laid in a uniform manner with adequate overlap. For maximum bond strength the overlap should be half the length of the brick.

The number of facing bricks for a given area increases with the number of headers and this may increase costs by a significant amount if expensive facings are specified. In half-brick thick cavity walling this cost factor may become important because of the introduction of headers cut from whole bricks.

Bonds requiring excessive cutting of units tend to be uneconomical in both materials and workmanship.

Complicated bond patterns may be acceptable for prestige buildings but significant additional costs are likely to be incurred where bond patterns require extra skill and care in the bricklaying process.

Good appearance with most bonds requires careful alignment of the perpend joints and if specified in conjunction with units of variable dimensions can result in additional costs. Problems can also arise when bricks of regular dimensions are used in conjunction with irregular units as a feature and a rigid coursing specification is imposed.

Bond types

Stretcher bond. This is the normal bond for walls of half-brick thickness (Figure 16.13) and is economical because of the limited need to cut bricks. Figure 16.12 shows the spread of load when the bricks are laid dry. However, when the units are bonded with mortar they tend to act as a composite and it is usual to assume the dispersal of load at 45°.

Stretcher bond with snap headers. This bond (Figure 16.14) is an easy way to improve the appearance of half-brick walling. A variety of patterns is possible.

Raking stretcher bond. Once again this is an economical and perhaps more interesting version than normal Stretcher bond (Figure 16.15). It is important to select the colour of the mortar carefully to avoid the joints becoming too prominent.

English bond. English bond (Figures 16.16 and 16.17) consists of one course of headers and one course of stretchers alternately. It produces a strong bond which is easy to lay but has a certain monotony of appearance. In this bond bricks are laid

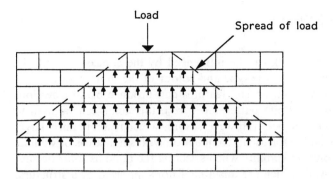

Figure 16.12 Spread of load in stretcher bond without mortar (see text)

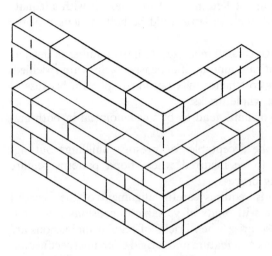

Figure 16.13 Brickwork – stretcher bond

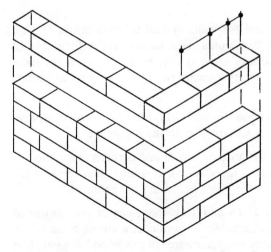

Figure 16.14 Brickwork – stretcher bond with snap headers

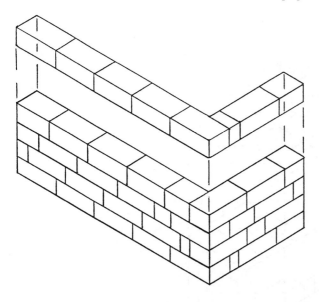

Figure 16.15   Brickwork – raking stretcher bond

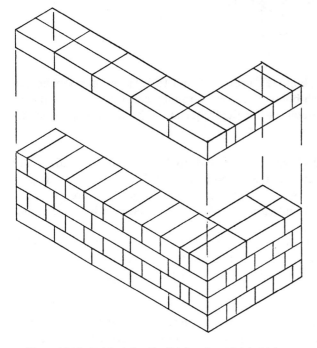

Figure 16.16   Brickwork – English bond one-brick thickness

as stretchers only on the boundaries of courses on the face of the wall and there is no break in the joints in a course running through the wall. The closer brick in the header course is always placed next to the quoin to complete the bond.

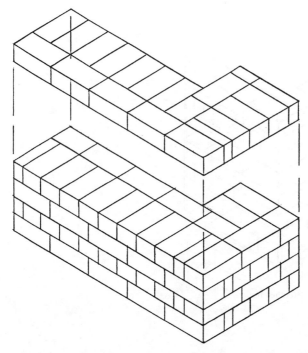

Figure 16.17   Brickwork – English bond one-and-half brick thickness

English garden wall bond. This bond (Figures 16.18 and 16.19) is more economical in facing bricks than true English bond. The bond normally consists of one course of headers to three courses of stretchers but the frequency of header courses is sometimes varied.

Flemish bond. Flemish bond (Figures 16.20 and 16.21) consists of headers and stretchers alternately in the same course to both front and rear elevations. It produces a simple pattern which is sometimes considered to give a more attractive appearance than English bond and easier to obtain a fair face on both elevations in walls of one brick thickness.

A variation of this bond sometimes called Single Flemish bond is used for walls of one-and-a-half brick thickness or more. It consists of Flemish bond on the face and English bond as backing. It is generally used where more expensive bricks are specified for the facing work.

Flemish Garden Wall bond (also known as Sussex bond). This bond (Figures 16.22 and 16.23) requires large areas of wall to show the pattern and perpend joints need to be kept true, especially if the headers differ from the stretchers in colour.

Flemish Cross bond. This bond (Figure 16.27) is similar to Flemish bond but uses two additional headers in place of a stretcher at intervals. It needs a large area of wall to be shown to advantage.

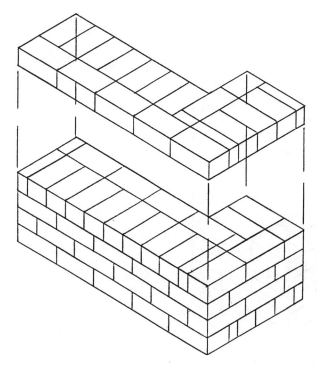

Figure 16.18   Brickwork – English garden wall bond one-and-half brick thickness

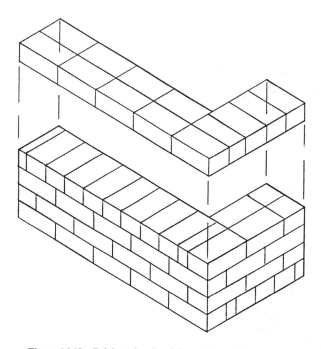

Figure 16.19   Brickwork – English garden wall bond one-brick thickness

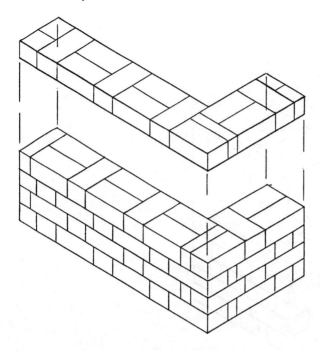

Figure 16.20   Brickwork – Flemish bond one-brick thickness

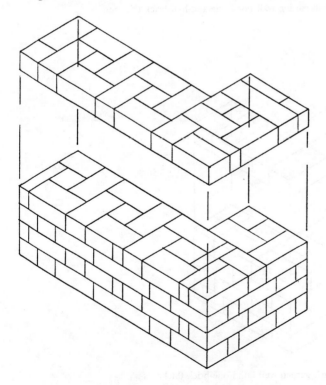

Figure 16.21   Brickwork – Flemish bond one-and-half brick thickness

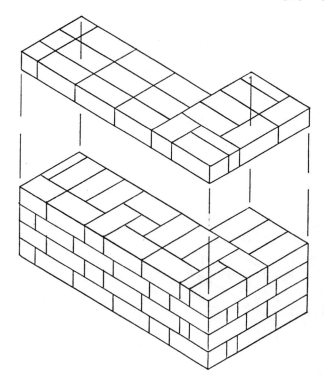

Figure 16.22 Brickwork – Flemish garden wall bond one-and-half brick thickness

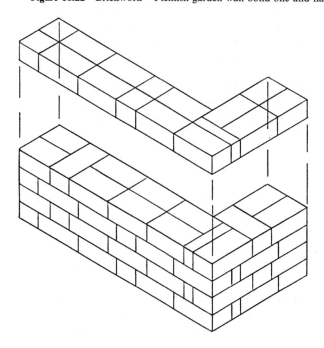

Figure 16.23 Brickwork – Flemish garden wall bond one-brick thickness

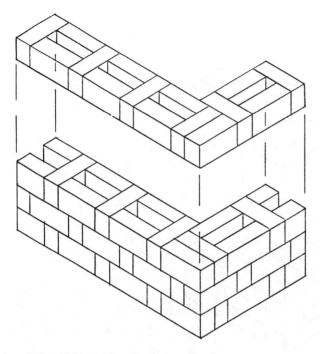

Figure 16.24   Brickwork – Rat-trap bond

Rat-trap bond. This bond (Figure 16.24) of brick-on-edge construction consists of headers and stretchers alternatively in the same course and produces an economical wall of unusual appearance but does not form a true cavity wall and consequently is unlikely to ensure resistance to rain penetration.

Quetta bond. Quetta bond (Figure 16.25) is similar to Flemish bond but in plan form the headers and stretchers form a 'T' rather than 'L' shape. It facilitates the introduction of vertical reinforcement in the voids, which are then filled with mortar. The reinforcement is usually taken into the foundations, floors and roof.

Monk bond. This bond (Figure 16.26) has two stretchers to one header in each course with headers staggered. It is a complicated bond to lay but gives an interesting appearance. Several variations of this bond are used.

Vertical stack bond. Stack bond (Figure 16.28) is used mainly for panel infills as it generally needs structural support. It is the quickest and most economical bond to lay as it eliminates the need for cutting bricks. The same comments apply to Horizontal Stack bond.

Basket weave bond. Figure 16.29 illustrates one of the many forms of Basket Weave bond. Various intricate bonds can be achieved using a mixture of horizontal and vertical bricks but like Stack bond it is usually used for infill panels in framed structures, as well as for non-loadbearing partitions.

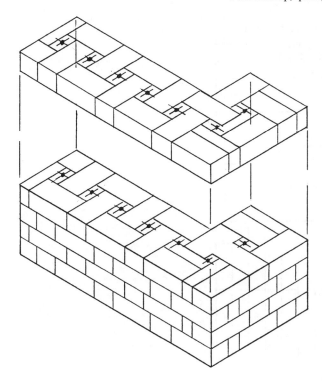

Figure 16.25   Brickwork – Quetta bond

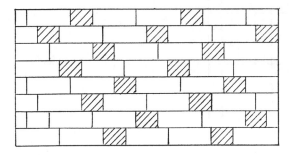

Figure 16.26   Brickwork – Monk bond

Herringbone bond. The bricks in this bond (Figure 16.30) are laid at an angle of 45°. It is normally used for walls four bricks or more in thickness and the units are laid commencing at the centre line of the wall and working towards the face bricks. Diagonal (laid at angles other than 45°) and herringbone patterns are often used to form ornamental panels in the face of walls, or for brick paving.

Header/brick-on-edge bond. This unusual bond (Figure 16.31) is formed using alternate courses of headers and brick-on-edge stretchers and gives a very strong horizontal emphasis.

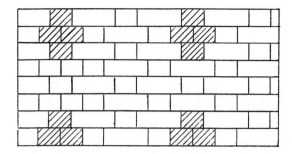

Figure 16.27   Brickwork – Flemish cross bond

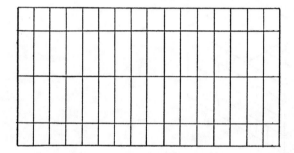

Figure 16.28   Brickwork – vertical stack bond

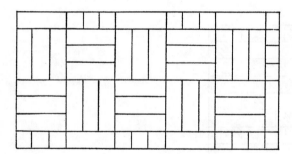

Figure 16.29   Brickwork – Basket weave bond

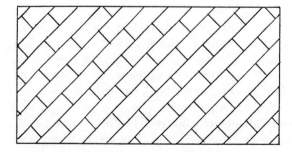

Figure 16.30   Brickwork – Herringbone bond

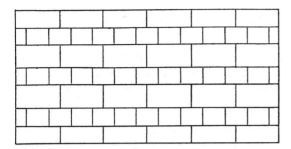

Figure 16.31   Brickwork – header/brick-on-edge bond

Heading bond. This bond (not illustrated) uses all the bricks as headers on the face and is used mainly for rounding curves.

English cross bond. This bond (Figure 16.32) is similar to English Bond but a bat is placed next to the end stretcher (from the left-hand end of Figure 16.32) every other course producing staggered stretchers as illustrated.

Dutch bond. This bond (Figure 16.33) is the same as English cross bond in the centre of the wall but without closures. Three-quarter bricks are used at the ends of the wall. Patterns or diapers can be picked out as illustrated.

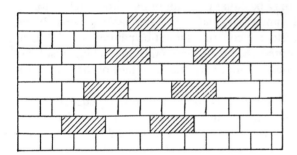

Figure 16.32   Brickwork – English cross bond

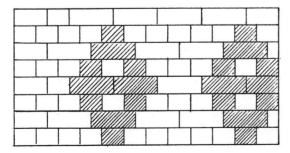

Figure 16.33   Brickwork – Dutch bond

16.4   Natural stone

Finishes

Ashlar (aisler). This type of stone is defined as square-hewn. Ashlar masonry normally consists of these finely dressed stones laid in courses with thin joints (5 mm approximately). Finishes are usually sawn and rubbed but tooled finishes can be supplied.

Rubble. This type of stone is generally roughly dressed with a hammer if any dressing is needed after removing from the quarry. The amount of dressing depends also on the type of construction and squared, coursed rubble will require a larger amount also squared, knapped flintwork.

Quoins usually offer a contrast to the remainder of the wall in the form of sawn or roughly squared blocks.

Types of walling

Ashlar walling
Due to its formal character this type of walling is generally appropriate to buildings in an urban setting. In solid walls, for reasons of economy, common brickwork or blockwork is generally used as a backing material and the stonework bonded or mechanically tied together.

The coursing of the stonework may be equal or random but should relate to the coursing of the backing to avoid unnecessary cutting of the backing units. If the stone is backed by structural concrete, the concrete should be painted with bitumen paint before any stone is fixed to protect the facing material from leaching of soluble salts which may have deleterious effects. Painting of the inner face of the stonework is not recommended.

When stone is used for ashlar cavity walls it is considered good practice to have a cavity of not less than 50 mm and not more than 75 mm (BS 5390: 1976). The stone may be backed with brick at alternate courses or consist of solid blocks of stone provided the wall satisfies the normal requirements of structural adequency.

Rubble walling
Due to its informal character walling of this type is more common in a village or rural setting. The term rubble walling covers a wide range of masonry and its characteristics depend largely on the region and the type of stone available locally. The types of rubble walling may be classified as follows.

Random rubble. This type of masonry is constructed of the stones as they come from the quarry. The stones are selected and placed in position to obtain a good bond with the minimum of cutting. For solid walls the transverse bond is achieved by the use of bonders (one per square metre minimum) which should be approximately two-thirds of the wall thickness; headers extending through the wall are known as 'through stones' but their use is not recommended due to the danger of moisture penetration.

Random rubble (coursed) This type of walling is similar to uncoursed random rubble except that the work is roughly levelled up to courses at intervals of 600–900 mm. The course heights usually correspond with the heights of the quoin and jamb stones.

Random rubble (shoddy work). This variant of wall type is built in graduating or diminishing courses, having stone of 200–225 mm at the base diminishing to 35–50 mm at the top.

Squared rubble (uncoursed). This type of walling consists of stones roughly squared as risers or jumpers and stretchers of varying heights and are laid uncoursed. The risers are generally not more than 250 mm in height and the stretchers approximately two-thirds the height of the adjacent riser. When small stones are used to assist bonding the masonry is known as 'sifted rubble'.

Snecked rubble. This consists of stones roughly squared but without the limitation of size and proportions associated with uncoursed squared rubble. It is designed to include a definite proportion of snecks, thus avoiding continuous vertical joints. The snecks are small stones but not less than 50 mm in any dimension.

Squared rubble (coursed). The work is levelled up to courses of varying depth from 300–900 mm using stones similar to those for snecked rubble. The courses usually correspond with the quoin or jamb stones.

Coursed walling
This consists of courses which may vary from 100–300 mm (225 mm average) but the stones are roughly squared in height in any one course. The faces may be dressed smooth or rock-faced. A chequered effect is sometimes produced by the introduction of smaller stones in the same course at intervals, these are known as 'pinnings'.

Polygonal rubble
This walling consists of stones with no pronounced lamination roughly pitched into irregular polygonal shapes and bedded to show the face joints running irregularly in all directions. If the stones are only roughly shaped and fitted it is known as 'rough-picked' but when the face edges are more carefully defined and fitted it is called 'close picked'. 'Kentish Rag' is another term for this type of walling when Kentish ragstone is used in irregular blocks.

Flint walling
This type of walling is traditional in East Anglia and South/South-east England. Flints or cobbles (popples) are used and these vary in size up to 300 mm in length and 75–150 mm in width and thickness.

These walls are built up with a facing of selected flints or cobbles and a core of the same material or other rubble. Quoins and dressings to openings are usually squared stone or brickwork. This type of work is only raised in short lifts to prevent the mortar squeezing out.

The flints may be whole or split and set in courses or uncoursed as required. When the flints are squared (knapped or snapped) the work is sometimes described as 'gauged' or 'squared' flint walling and the joints on the exposed face are well raked back.

Bonding/lacing courses of long thin stones, tiles or bricks are usually introduced at vertical intervals of 1–2 m and stone or brick piers (flush with the wall faces or projecting) at about 2 m centres.

Lakeland walling

This type of walling averages from 525–750 mm in thickness and is constructed of irregular flat-bedded slate blocks up to 600 mm in width, 1 m length and about 75 mm on bed. The walls consist of two well-bonded faces with a core and the slates are laid tilted downwards towards the outer face at a slope of approximately 1 in 6. Through-stones or 'watershots' are introduced at regular intervals and the slates roughly coursed. Bedding mortar is kept back about 50 mm from the face and the core is dry bedded. Quoins when used, are usually hammer-dressed, angle-drafted limestone or large sawn slate blocks.

Dry stone walling

This type of walling consists of roughly dressed stones laid on bed in dry earth or on edge at a slope and the core is formed of pise (compressed earth or stiff clay) or small stones. The stones are laid in such a manner that rain penetrating the outer face tends to run out at a lower level.

Surface treatments

Water repellents

BS 6477 (1984) 'Water repellents for masonry surfaces' specifies performance requirements, including requirements for durability, for water repellents intended for use on masonry above ground level and free from cracks exceeding 0.15 mm in surface width.

Masonry, when properly designed and constructed, provides adequate protection from the weather but colourless water repellent treatments based on solutions of waxes and/or metallic soaps have long been used in building primarily for remedial purposes. However, water repellents will not cure all problems of rain penetration, particularly if there is a basic fault in the detailing of the walls and they may sometimes be ineffective or even detrimental.

Modern durable treatments are usually based on silicone resins and polyoxo-aluminium stearate. BS 6477 (1984) classifies water repellents into four groups, three of which are appropriate to stone masonry as follows:

Group 1. Substrate predominantly siliceous, e.g. sandstones such as Darley Dale, Wealden, St Bees, Blaxter and Bollington stone.

Group 2. Substrate predominantly calcareous, e.g. natural limestone such as Portland, Bath, Hopton Wood, Clipsham, Weldon, Doulting, Caen. Cast stone made using hammer compacted factory processes.

Group 3. Freshly made cementitious materials and other materials of similar alkaline nature, e.g. new or repaired surfaces of stonework bonded, painted or rendered with cement-based materials.

On sandstone, the surface water repellence caused by silicone treatment is relatively long lasting but has no appreciable preservative effect. Treatment of poorer quality sandstone may cause flaking of the surface if the stone is contaminated with aggressive soluble salt such as sodium or magnesium sulphate.

When limestone is treated initially there is a retardation of the solution of the stone by acid rainwater, but the effect is transient and after a few months the surface repellence disappears and the rate of loss from wetted surfaces may actually increase. When limestone is contaminated with aggressive salts flaking of the treated surface may occur. Subsequent treatments need to be frequent to prevent any surface loss and may increase the rate of decay when the stone is contaminated with salts.

The life of the treatment is improved if the repellent contains silicon ester and friable surfaces may become somewhat consolidated. However, when aggressive salts are present the risk of spalling of the treated surface is increased.

The performance of water repellents is influenced by location, exposure and the position of the stone in the building. Proprietary stone preservatives are available and attempts to develop new, deeply penetrating materials are in progress. These materials are not currently included in a British Standard and should not be confused with water repellents.

New stone should not require treatment with these materials if the materials have been carefully selected and the work carried out in accordance with the appropriate British Standards.

References

British Ceramic Research Limited. (1988) *Model specification for Clay and Calcium Silicate Structural Brickwork.* SP56.
BS 5628: Part 1: 1978 *Structural use of unreinforced masonry.* BSI, London.
BS 5628: Part 2: 1985 *Structural use of reinforced and pre-stressed masonry.* BSI, London.
BS 5628: Part 3: 1985 *Use of masonry, Materials and components, design and workmanship.* BSI, London.
BS 5390: 1976 *Stone masonry.* BSI, London.
BS 6477: 1984 *Water repellents for masonry surfaces.* BSI, London.
Buildings Research Establishment Digest No. 71 (1st Series), (1956) *Frog up or frog down.*

Repairing and replacing masonry

17.1 General considerations

Before any remedial works are carried out it is necessary to know what caused the damage. However, diagnosis is not always easy and frequently can be attributed to more than one cause.

Cracking of masonry is perhaps the most common defect and until the causes have been identified there can be no certainty that a simple repair of the cracks is all that is needed. When serious damage occurs the nature of the remedial work will depend upon whether or not the stability of the walls or structure has been affected. This chapter considers defects arising from the following causes and methods of repair are outlined:

(1)   frost action
(2)   sulphate attack on mortar and rendering
(3)   the use of unsound materials
(4)   efflorescence
(5)   corrosion of embedded metal
(6)   cracking due to thermal and moisture movements
(7)   ground movements
(8)   fire
(9)   roof spread

17.2 Frost action

In the UK frost failures of masonry are generally confined either to new work or work which, because of the conditions of exposure, remain wet during the winter.

The characteristic effects of frost attack on extruded clay bricks and sedimentary type stones is delamination (or, if very severe, more general disintegration) and erosion of the mortar joints. Pressed-clay bricks and calcium silicate bricks, when attacked tend to crumble or show signs of general erosion. Concrete bricks and blocks generally have good frost resistance but when damage occurs this tends to cause surface scaling. In some instances, frost may not have been the primary cause of damage (e.g. the initial defect may have been sulphate attack) and may only have exacerbated the initial defect.

Diagnosis
In order to establish that frost action is the only cause of the defects it is necessary to eliminate the other possible causes of similar defects. If sulphate attack is also suspected confirmatory laboratory tests may be necessary. Chapters 5 and 6 discuss frost attack and sulphate attack respectively in more detail. Frost failures are usually confined to partly-built unprotected masonry or to masonry subject to conditions of 'severe' exposure, i.e. free-standing walls, parapets and retaining walls and occasionally masonry between ground level and the d.p.c. Units intended for use in such locations should be selected for their known properties of good frost resistance (see Chapter 3).

Remedial action
When the mortar has been damaged without disturbing the masonry the joints should be raked out to a depth of at least 16 mm and re-pointed. The pointing mortar should be compatible with the units. A common mistake is to use too strong a pointing mortar. The re-pointing should preferably be carried out in spring or summer (particularly if it is of low strength). Alternatively, the wall should be protected by coverings until the work has hardened (see Chapter 5).

Spalling of the units usually affects only a small proportion of those in the wall. If the remainder appear sound, the damaged units can be cut out and replaced by sound units having a sufficiently high frost resistance to withstand the conditions of exposure.

If frost damage is extensive it may be cheaper and preferable to re-build it entirely. Alternatively, if rendering is an acceptable solution to the problem, all loose particles of the existing wall should be removed and the wall rendered on stainless steel (or other suitably protected) mesh or expanded metal fixed to the wall.

When frost damage is extensive below d.p.c. level in free-standing walls it is not generally possible to cut out and replace the units without affecting the stability of the wall and it may be necessary to re-build the wall. Alternatively, it may be more acceptable to repair the wall with a modified coloured mortar, e.g. a specially prepared Styrene Butadiene Resin mortar (SBR) containing a white Portland cement and appropriate pigment to blend with the existing units. The manufacturer of the SBR should always be consulted regarding the suitability of their product for a specific location, as well as the mix proportions and general procedure.

17.3 Sulphate attack

Sulphate attack on mortar and renderings generally only occur in association with clay masonry. Exceptions could be in polluted industrial atmospheres or when masonry is buried or retains sulphate-bearing soils. More detailed information is given in Chapter 6.

The defects caused may appear in various forms:

Unrendered masonry

(a)   the mortar expands leading to deformation and cracking of the masonry;
(b)   the edges of individual units spalls;
(c)   the mortar deteriorates commencing with surface effects similar to those caused by frost, followed by lamination of the joints and ultimately resulting in serious disintegration of the masonry.

Rendered masonry

(a)   crazing of the rendering occurs which may originally have been caused by shrinkage of the rendering as it dries, accentuated and followed by the formation of horizontal cracking along the mortar joints;
(b)   wide horizontal and vertical cracks appear in the rendering, some outward curling edges may also be observed at the cracks;
(c)   adhesion of the rendering to the backing fails so that areas of rendering become detached. Often the exposed masonry shows signs of efflorescence.

Chimney stacks. In addition to the signs mentioned above, sulphate attack in chimney stacks may produce curvature or leaning of the stack.

Diagnosis
The signs mentioned above are frequently sufficient to enable failures from sulphate attack on mortars and renderings to be recognized. If diagnosis is difficult or a dispute occurs, chemical analysis of samples of mortar or rendering will produce the necessary confirmatory evidence.

Remedial action
Once sulphate attack has been diagnosed it is usually too late to carry out effective remedial action, as the structural integrity of the masonry is almost certainly impaired. If sulphate failure is detected in its early stages and there is unlikely to be any danger, due to the reduced structural integrity of the walls, the following remedial action could be taken:

(1)   The sources of moisture should be located and measures taken to eliminate further entry into the masonry.
(2)   The defective masonry should be re-pointed using sulphate resisting cement.
(3)   All cracked and non-adherent rendering should be removed, the construction allowed to dry, all loose material and efflorescence removed and the wall re-rendered again using sulphate-resisting cement.

17.4   Use of unsound materials

Unsound materials may show one or more of the following characteristics:

(a)   pitting of mortar joints with modules of friable material in the nucleus of the defect;
(b)   strong pointing mortars are displaced and pits develop in the weaker bedding mortar;
(c)   general expansion occurs, with deformation and subsequent cracking of the masonry; and
(d)   individual clay bricks containing nodules of lime burst at the surface of the unit.

Diagnosis
Unsound materials cause the masonry to expand in a manner similar to that caused by sulphate attack except that lamination of the mortar joints is unlikely to occur. The defect usually occurs when unsound lime has been used in the mortar.

If large nodules of lime are present adjacent to the surface of clay bricks and the lump of lime is allowed to slake slowly by absorbing moisture from the air it is liable to expand and cause splitting of the brick face which can be extremely unsightly.

Remedial action
If unsound lime is present in the mortar joints the action ceases after hydration, and if measures are taken to exclude moisture from the masonry, repairs can generally be confined to any necessary repointing. With grossly unsound mortar, rebuilding is the only satisfactory remedial action.

Damaged individual clay bricks can either be replaced or the areas where particles of expanded lime have caused spalling can be repaired with a coloured mortar using a suitable pigment or brick dust as an aggregate.

17.5   Efflorescence

The most common effect of soluble salts in masonry is to produce an efflorescence which, although unsightly, is usually temporary and harmless. Efflorescence can occur on the units and/or the mortar joints (see Chapter 6). When it occurs on the units it is usually more noticeable on clay units than on natural stone, concrete or calcium silicate units.

The following are the most common defects arising from crystallization of water-soluble salts in masonry:

(a)   the appearance of a white deposit (efflorescence) on the surface of the masonry. This frequently appears during the first spring and summer after building;
(b)   decay of individual units in the exposed face.

Diagnosis
The appearance of efflorescence is well known (it should not be confused with leaching of free lime from concrete or mortar joints (see Chapter 6) and needs no discussion except to mention that in cases where the deposit is heavy an analysis of

the salts is often advisable in order to determine whether the salt is of a kind which may be expected to lead to deterioration.

Remedial action
Removal of efflorescence has been discussed in Chapter 6.

Where odd units have disintegrated due to cryptoefflorescence, they should be cut out and renewed with matching units having a low soluble sulphate content. If general disintegration of the wall face has occurred, the only satisfactory treatment is to render the surface or, in extreme cases, to re-build the wall.

17.6   Corrosion of embedded metal

The corrosion of metal when embedded or enclosed in masonry may produce the following defects:

(a)   opening of the mortar joints near the enclosed metal;
(b)   staining of the masonry due to rust;
(c)   cracking of the masonry where the partially embedded metal enters the wall.

Diagnosis
Items (b) and (c) above are usually sufficiently visible signs to diagnose the cause but with item (a) the cracking might be confused with that due to other causes. If rust stains are not apparent, the best method of investigation is to open up the masonry near the crack.

Cavity wall tie corrosion is now a well-known problem and is generally most apparent when inadequately galvanized vertical twist wall ties corrode and expand causing cracking of the masonry. A typical failure pattern is horizontal cracking in the external face of a wall every 300, 375 or 450 mm. However, walls which begin to crack are often re-pointed, effectively disguising the problem. The investigator should look out for excessively large mortar beds at regular vertical intervals. Also in gable walls, the ridge may have been lifted giving what is now commonly known as the 'pagoda effect'. The corrosion of the lighter-gauge butterfly wire ties can have a more insidious effect in that as the expansion of the embedded metal may not be sufficient to disrupt the mortar joints the first sign of trouble could be when the outer leaf of the wall collapses.

Reason for defects
Unprotected or inadequately protected ferrous metal will eventually corrode when in contact with moisture and air but the corrosion is more rapid in the presence of acids, sulphates and chlorides. Polluted industrial environments contain the first two and sea spray the latter two of these corrosive agents.

Corrosion is most likely to occur when the metal is in direct contact with external masonry or partially embedded in such masonry. The rust formed by corrosion of steel occupies a volume as much as seven times greater than the original metal, so generating expansive forces in the mortar bed.

Remedial action
When the corrosion of metal is detected before the masonry has suffered serious damage, measures can be taken to inhibit it or reduce the rate of further corrosion as follows:

(i) Cut away the masonry, if necessary in sections, to expose the metal.
(ii) Clean the latter thoroughly, prime with a rust-inhibitive primer and coat with a bitumen paint or similar.
(iii) Alternatively clean the metal and case in dense concrete or a modified mortar, e.g. SBR.

If only the end of the metal is embedded, this should be replaced, if possible, with an austenitic stainless steel component. If it is not possible to replace the original metal, particular attention should be paid to the protection of the metal where it enters the masonry and the socket should be caulked with a suitable mastic at the point of entry.

When light-gauge wall ties corrode they rarely damage the mortar joints because of the small amount of expansion due to corrosion and it is usually sufficient to replace them with one of the many proprietary replacement-type wall ties. However, when vertical twist-type wall ties corrode it is necessary to take more drastic remedial action as follows:

(a) The ties should be located using a metal detector and the positions marked on the face of the masonry.
(b) The masonry units should then be removed above each tie, usually with a power chisel. It is important that the units are removed in a pattern and manner which will not affect the stability of the wall. If the units are carefully removed they can be re-used, saving expense and the problem of colour and size matching.
(c) The ties should then be removed after loosening by drilling each side of the mortar joint in the inner leaf. Alternatively, specially designed hydraulic tools are sometimes used to extract the ties.
(d) Existing mortar should then be thoroughly cleaned from the brick on which the new tie is to be bedded and the proprietary tie is then secured to the inner leaf.
(e) The units should then be carefully re-built into the work. It is recommended that on completion of the above work that the whole area is re-pointed if only for aesthetic reasons, the joints having been chased out to the correct depth at the time the units were removed before removal of the wall ties.

17.7 Cracking due to thermal and moisture movements

Thermal and moisture expansion

If provision for thermal and moisture expansion is not made in long lengths of clay masonry or if due account is not taken of the geometry and other features as described in Chapter 4 then one or more of the following defects may occur:

(a)   cracking of the masonry;
(b)   oversailing of the masonry at the ground floor d.p.c. level;
(c)   bowing of the masonry between vertical or horizontal supports or returns;
(d)   spalling of the masonry at points of restraint;
(e)   instability of the walls.

Drying shrinkage

Concrete and calcium silicate masonry tends to shrink rather than expand and allowance needs to be made for this movement in long walls and/or because of the geometry/special features such as openings and changes in section (see Chapter 4) if cracking of the masonry and instability of the walls is to be avoided.

Diagnosis

It is important to establish the reason for cracking before any remedial work is carried out. If it is due to thermal and moisture movements relatively simple remedial work may be sufficient. However, if the cause is structural and/or because of ground movements more drastic measures may be necessary. Chapter 4 outlines many of the defects associated with dimensional changes in masonry and these should be considered individually. Frequently when cracking occurs it is due to more than one cause.

Typical signs of thermal/moisture expansion in clay masonry are oversailing at the ground floor d.p.c. and vertical cracking in the walls, this is because of the fluctuating movements above d.p.c. level and the more constant moisture conditions below d.p.c. level, as well as the restraint provided by the foundations.

Damage due to drying shrinkage in free-standing walls tends to occur as vertical cracks at regular intervals whereas cracking under long low windows is frequently diagonal at each end in a 'back of an envelope' type crack pattern.

BRE Digest 251 (1981) classifies damage under three headings 'aesthetic', 'serviceability' and 'stability'. The first comprises damage which affects only appearance of the building. The second includes cracking and distortion which impairs the weather tightness or other function of the wall (e.g. sound insulation of a party wall may be degraded), fracturing of service pipes and jamming of doors and windows. In the third category are cases where there is an unacceptable risk that some part of the structure will collapse unless preventive action is taken.

Remedial action

If cracking is due to thermal and moisture movements, movement or control joints should be cut in the masonry at the centres and in the manner described in Chapter 4. Stability of the walls must be considered before cutting the joints (or provision of vertical slip planes at returns) and it may be necessary to provide replacement type wall ties and/or additional support for the walls.

Cracking of the masonry may be diagonal following the mortar joints or vertical passing through alternate joints and the intervening units. In remedial work this distinction is significant only when it is considered necessary to cut out and replace the cracked units. The decision whether or not to repair cracks and cut out units will depend mainly on two considerations: (a) whether the cracks are thought to be unsightly if not repaired; and (b) they are likely to encourage rain penetration.

Whether or not a crack is unsightly is highly subjective; all masonry walls contain cracks but usually these are hair cracks (up to 1 mm) which have no significance from either a structural or durability point of view. BRE Digest 251 (1981) defines cracks as negligible (hair-line), very slight, slight, moderate, severe and very severe. For stronger mortar joints, either cut out the units adjacent to the crack and rebond with a mortar designation (iii). If cracks pass through the units and mortar, cut out and rebond, using mortar similar to that in the existing wall. When repairing masonry as suggested above it is important to ensure good adhesion between the new units, and also not to use too strong a mortar otherwise shrinkage of the new mortar may cause a fresh crack to develop. If high-suction clay bricks are used these may need to be wetted before the mortar is applied; concrete and calcium silicate units should not be wetted.

If bowing of masonry has occurred between vertical or horizontal supports to an unacceptable degree re-building of the work is usually the only satisfactory solution to the problem. When spalling of masonry occurs at points of restraint the remedial action will be similar to that described under 'Frost action' on p. 298.

17.8 Ground movements

Masonry may be affected by ground movements due to consolidation settlement resulting from construction loading and movements which take place independently of imposed loading as described in Chapter 4 (see also BDA Design Note 1, Al-Hashimi, 1977). When cracking of masonry is due to ground movements it is important to bring in an expert.

Diagnosis
It is essential to carry out a detailed survey and this should be recorded photographically and by plotting the damage and distortion of the structure. Whenever possible cracks should be identified as either tensile or compressive cracks, i.e. compressive cracks are indicated by small flakes of masonry squeezed from the surface by localized crushing. Similarly, shear cracks are indicated by relative movement along a crack at points on the opposite side of it. Generally, cracks produced by foundation movement are not widely distributed throughout a building but tend to be concentrated in areas where maximum structural distortion and structural weak points coincide.

When foundation movement occurs cracks usually show externally and internally and may extend through the d.p.c. and down into the foundation. It is often possible to determine whether the foundations are sagging or hogging by the taper of the cracks in the masonry and a thorough examination of the foundation in the area of most movement should be carried out and the nature of the underlying ground assessed. It is also important to determine if ground movements are progressive and this is usually carried out using vernier type tell-tales fixed to the masonry.

Displacement and wedging of windows and doors is also a useful indicator of local settlement. For more detailed information readers are referred to BRE Digest 251 (1981).

Remedial action

The cause of the movement should first be established and the advice of an expert in geotechnic problems sought.

If it is established that the damage has been caused due to changes in moisture content of the soil, underpinning of the walls is usually the answer for minor damage or the provision of an extra foundation would take the load to relatively stable ground below the zone of seasonal moisture changes (Al-Hashimi, 1977). Cracks in the masonry can then be repaired as described earlier in this chapter.

17.9   Fire

The effects of fire vary dependent upon the severity of the fire, the thickness of the walls and the type of units. In a severe fire cracking and bulging of the masonry may occur in addition to spalling of the surface. Clay bricks having been exposed to fire in the manufacturing process are generally less vulnerable than concrete or calcium silicate units. However, sudden quenching of masonry with water in the course of fire-fighting may cause spalling, some perforated units with holes near the surface being particularly susceptible to such damage.

The most serious damage to masonry is usually caused by the distortion of built-in structural members caused by thermal movement at high temperatures.

Diagnosis

The effects of fire on masonry walls is usually self evident as described above. However, when mortar or concrete is exposed to temperatures in excess of 500°C general weakening and friability occurs as described in Chapter 15. The temperatures the wall has been subjected to can be assessed by examining colour changes as described in Chapter 15. This may not be evident purely by examination of the wall surface and mortar samples should be taken for this purpose.

It is also suggested that when cavity walls are subjected to major fires that some check on the wall ties is necessary before passing a wall as sound. If the ties are galvanized they may have been damaged as zinc vaporizes at approximately 400°C.

Similarly, if polypropylene wall ties have been used these should be checked to ensure that they have not deteriorated. Damp proof courses and other built-in components should be checked after fires as these may be damaged and/or rendered ineffective.

Remedial action

Cracking and spalling can generally be repaired as described earlier in this chapter.

If light-gauge butterfly wall ties or polypropylene wall ties are damaged it is usually sufficient to install replacement type wall ties. However, if vertical twist wall ties have deteriorated it is advisable to remove and replace them as described on p. 309. After severe fires analysis of mortar samples as described in Chapter 15 will reveal if re-pointing of the wall is sufficient or if re-building of all or part of the wall is advisable.

17.10   Roof spread

Pitched roofs that have been inadequately tied (or the members cut or removed for services) will, under the action of dead, snow and wind loads, spread appreciably at the eaves and displace the masonry.

Diagnosis
Displacement of the wall and frequently lintels over windows in the region of the eaves will often be the first sign of this problem and subsequent examination of the roof space invariably confirms the diagnosis.

Remedial action
The roof structure must be suitably strengthened to remove the outward thrust at or near the top of the wall.

Re-building of the affected courses of masonry is usually the only satisfactory solution to the problem.

17.11   Replacing defective wall ties

When wall ties corrode, or it is discovered that an insufficient quantity of ties have been provided, the usual solution to the problem is to install one of the many types of replacement wall tie. However, the first task is to have an independent survey carried out to establish the condition of the cavity walls and the existing wall ties. If the wall ties are of light gauge it is generally satisfactory to leave them in position but if vertical twist-type wall ties (i.e. strip ties of over 3 mm thickness) have corroded it will be necessary to remove them as (previously stated on p.300) expansion of the corroded metal is capable of generating expansive forces in the mortar beds (Moore, 1981). If vertical twist-type ties have corroded this is usually evident where the mortar joints have cracked or where some of the horizontal joints are wider than normal and have been re-pointed.

Unless the surveyor is using an electromagnetic metal detector capable of being calibrated to assess the embedment of the wall ties it will be necessary to check the actual embedment using a Borescope and/or by cutting out individual bricks. Specialist firms offering a wall tie survey service should be able to demonstrate that their surveying techniques include an assessment of depth of embedment. The mere use of a metal detector is not sufficient as all this does is locate metal some-where between the external face of the wall and the inner leaf. More detailed information on corrosion of steel wall ties: recognition, assessment and appropriate action is available from the BRE (de Vekey, 1979a).

If the stability of one or both of the leaves of the cavity wall has been affected, or where obvious damage has occurred such as bulging or splitting the only satisfactory remedial action may be to rebuild one or both leaves of the wall; however, always first consult a qualified structural engineer.

On the assumption that both leaves of the wall are basically sound and that rusty vertical strip cavity wall ties have been found these must be removed.

Replacement wall ties with double-ended expander fixings
The installation of this type of tie requires only a masonry drill and an appropriate torque spanner/device. The ties are normally lockable at each end so that the inner fixing can be site tested before the fixing in the outer leaf is locked. The screw/core device needs to be tightened to a predetermined torque (this value is provided by the manufacturer) and some ties have a shaft with a local reduction in diameter and on tightening the shaft fractures when the tie is subjected to its design torque.

The method of installation varies dependent upon the type of anchor but basically the operation involves:

(a)   selecting the correct length of tie for the purpose;
(b)   drilling through the outer leaf and into the inner leaf to a predetermined depth;
(c)   cleaning out the hole;
(d)   inserting the tie into the hole and tightening it. Site testing can then be carried out to ensure an adequate strength and manufacturer's are usually willing to provide this service;
(e)   inserting the cone device in the outer leaf and tightening to the required torque;
(f)   when all the ties are installed the holes in the bricks are made good with a mortar or resin to blend with the bricks.

Manufacturers normally recommend that the replacement ties are positioned in the centre of a brick but sometimes they will agree to the ties being installed in the mortar bed joints. Problems can arise if the inner leaf is hollow/cellular blockwork or deep frogged bricks have been laid frog-down. The same problem arises if the external leaf of the wall is constructed in some types of perforated brick. If such difficulties do arise then either an alternate type of fixing is required or the voids will need to be filled with grout or resin before fixing of the anchors takes place.

Most of these ties have a washer in the centre to form a rainwater drip, but if this is not so they should slope down towards the outer leaf.

Resin-grouted anchors
These ties should generally only be installed in dry walls if both ends of the anchors rely on resin grout for fixing. However, some anchors are fixed with resin-grout in the inner leaf only and an expander fixing in the external leaf.

The procedure for fixing is similar to that for double-ended expander ties if the external fixing is mechanical, but for the inner leaf after drilling is complete a thin-walled tube is inserted and a capsule of adhesive inserted in the hole in the inner leaf via the tube using a push rod. The tie is then usually forced into the inner leaf using a drill with a special attachment so that the capsule is fractured and the two-part resin adhesive suitably mixed. The resin is then allowed to set hard (most resins achieve this in approximately 1 hour dependent upon temperature) (de Vekey, 1979b) and then the leaf is secured as for double-ended expander ties or by using an adhesive.

High-strength foam
If it is considered appropriate to fill the cavity a high-strength polyurethane foam can be used. This has the advantage of increasing the thermal insulation of the wall but the long-term durability of its adherence is unknown.

Numerous combinations of double-ended expander ties and resin-grouted anchors are possible and the use of plastic expanders (PVC) has been found to be effective particularly in the weaker bricks and blocks.

Special spiral rod type ties. These have been used with adhesives and some varieties can be used in conjunction with timber frame construction (BRE Digest, 1982).

Modifications
Modifications of traditional ties which can be plug-fixed to the inner leaf can be used if bricks are removed from the outer leaf.

Whatever the type of replacement anchor chosen it is vital that it should be corrosion-resistant i.e. austenitic stainless steel or non-ferrous metal.

Fixing of replacement type anchors should only be performed by supervised and well-trained staff; in addition the work ideally should be carried out only by specialist contractors.

A range of remedial-type wall ties is illustrated in Figures 17.1–17.3.

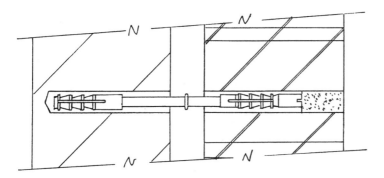

Figure 17.1   Double expander fixing

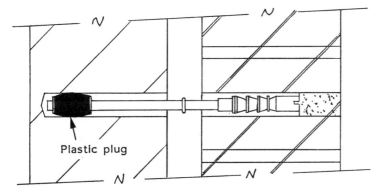

Figure 17.2   Expander/plastic fixing

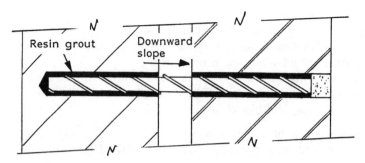

Figure 17.3 Resin fixing.

References

Al-Hashimi, K. (1977) *Foundation design: soil/structure interaction in shrinkable soils.* BDA Design Note 1.

BRE Digest 251 (1981) *Assessment of damage in low-rise buildings.* July. Buildings Research Establishment.

BRE Digest No. 257 (1982) *Installation of wall ties in existing construction.* January.

de Vekey, R. C. (1979a) *Corrosion of steel wall ties: recognition, assessment and appropriate action.* BRE Information Paper 28/79, October.

de Vekey, R. C. (1979b) *Replacement of cavity wall ties using resin-grouted stainless steel rods.* BRE Information Paper 29/79,October .

Moore, J. F. A. (1981) *The performance of cavity wall ties.* BRE Current Paper 3/81.

Index